Introduction to Geopolitics

This lucid and concise introductory textbook guides students through their first engagement with geopolitics. It offers a clear framework for understanding contemporary conflicts by showing how geography provides opportunities and limits upon the actions of countries, national groups, and terrorist organizations. The overarching theme of geopolitical structures (contexts) and geopolitical agents (countries, national groups, individuals, etc.) is accessible and requires no previous knowledge of theory or current affairs.

Throughout the book, case studies including the rise of al-Qaeda, the Korean conflict, Israel–Palestine, Chechnya, and Kashmir emphasize the multi-faceted nature of conflict. These, along with activities, aid *Introduction to Geopolitics* in explaining contemporary global power struggles, the global military actions of the United States, the persistence of nationalist conflicts, the changing role of borders, and the new geopolitics of terrorism. Throughout, the readers are introduced to different theoretical perspectives, including feminist contributions, as both the practice and representation of geopolitics are discussed.

Introduction to Geopolitics is extensively illustrated with diagrams, maps, and photographs. Reading this book will provide a deeper and critical understanding of current affairs and facilitate access to higher level course work and essays on geopolitics. Both students of geography and international relations and general readers alike will find this book an essential stepping-stone to a fuller understanding of contemporary conflicts.

Colin Flint is an Associate Professor of Geography at the University of Illinois. His research interests include geopolitics and hate groups. He is editor of *The Geography of War and Peace* (2005) and *Spaces of Hate* (2004) and co-author, with Peter Taylor, of *Political Geography: World-Economy, Nation-State, and Locality* (2000).

Introduction to Geopolitics

Colin Flint

Routledge
Taylor & Francis Group

LONDON AND NEW YORK

First published 2006
by Routledge
2 Park Square, Milton Park, Abingdon, Oxon OX14 4RN

Simultaneously published in the USA and Canada
by Routledge
270 Madison Ave, New York, NY 10016

*Routledge is an imprint of the Taylor & Francis Group,
an informa business*

Transferred to Digital Printing 2010

© 2006 Colin Flint

Typeset in Times New Roman by
Florence Production Ltd, Stoodleigh, Devon

British Library Cataloguing in Publication Data
A catalogue record for this book is available from the
British Library

Library of Congress Cataloging in Publication Data
Flint, Colin (Colin Robert)
   Introduction to geopolitics / Colin Flint.
        p. cm.
   Includes bibliographical references and index.
   1. Geopolitics—Textbooks.   I. Title.
   JC319.F55 2006
   320.1'2—dc22   2005033530

ISBN10: 0–415–34494–8 (hbk)
ISBN10: 0–415–34493–X (pbk)

ISBN13: 978–0–415–34494–4 (hbk)
ISBN13: 978–0–415–34493–7 (pbk)

CONTENTS

FIGURES

TABLES

BOXES

ACKNOWLEDGMENTS

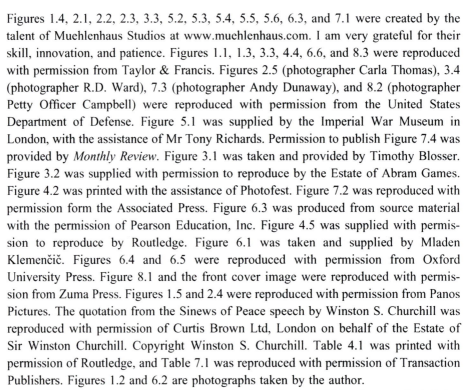

Figures 1.4, 2.1, 2.2, 2.3, 3.3, 5.2, 5.3, 5.4, 5.5, 5.6, 6.3, and 7.1 were created by the talent of Muehlenhaus Studios at www.muehlenhaus.com. I am very grateful for their skill, innovation, and patience. Figures 1.1, 1.3, 3.3, 4.4, 6.6, and 8.3 were reproduced with permission from Taylor & Francis. Figures 2.5 (photographer Carla Thomas), 3.4 (photographer R.D. Ward), 7.3 (photographer Andy Dunaway), and 8.2 (photographer Petty Officer Campbell) were reproduced with permission from the United States Department of Defense. Figure 5.1 was supplied by the Imperial War Museum in London, with the assistance of Mr Tony Richards. Permission to publish Figure 7.4 was provided by *Monthly Review*. Figure 3.1 was taken and provided by Timothy Blosser. Figure 3.2 was supplied with permission to reproduce by the Estate of Abram Games. Figure 4.2 was printed with the assistance of Photofest. Figure 7.2 was reproduced with permission form the Associated Press. Figure 6.3 was produced from source material with the permission of Pearson Education, Inc. Figure 4.5 was supplied with permission to reproduce by Routledge. Figure 6.1 was taken and supplied by Mladen Klemenčič. Figures 6.4 and 6.5 were reproduced with permission from Oxford University Press. Figure 8.1 and the front cover image were reproduced with permission from Zuma Press. Figures 1.5 and 2.4 were reproduced with permission from Panos Pictures. The quotation from the Sinews of Peace speech by Winston S. Churchill was reproduced with permission of Curtis Brown Ltd, London on behalf of the Estate of Sir Winston Churchill. Copyright Winston S. Churchill. Table 4.1 was printed with permission of Routledge, and Table 7.1 was reproduced with permission of Transaction Publishers. Figures 1.2 and 6.2 are photographs taken by the author.

Thanks to Jessica Schmid, Joel Chorny, Jessica Palmer, and Matt Bestoso for helping me to find and collect data. The diligence and hard work of Zoe Kruze at Routledge was essential for the successful completion of this project. I also thank Andrew Mould for his editorial guidance and for suggesting the book in the first place.

Last, but by no means least, thank you to all the students from the many classes of GEOG 128 *The Geography of International Affairs* I had the pleasure and privilege to teach during my time at Penn State. You were the inspiration and motivation for this book.

ABBREVIATIONS

ASSR	Autonomous Soviet Socialist Republic
BJP	Bharatiya Janata Party
CSTO	Collective Security Treaty Organization
DARPA	Defense Advanced Research Projects Agency
GEOINT	Geospatial Intelligence
HUM	*Hizbul Mujahideen*
IMET	International Military Education and Training Program
IMF	International Monetary Fund
IRA	Irish Republican Army
ISI	Inter-Services Intelligence
ISL	Islamic Students League
JEI	*Jamaat-e-Islami*
JKJEI	*J & K Jamaat-e-Islami*
JKLF	Jammu and Kashmir Liberation Front
JKPC	Jammu and Kashmir People's Conference
KDP	Korean Democratic Party
MAD	mutually assured destruction
NAFTA	North American Free Trade Agreement
NATO	North Atlantic Treaty Organization
NGA	National Geospatial-Intelligence Agency
NGO	non-governmental organization
NSC	National Security Council
NSG	National System for Geospatial-Intelligence
NSS	National Security Strategy
PLO	Palestine Liberation Organization
POTB	Prevention of Terrorism Bill
SWNCC	State-War-Navy Coordinating Committee
UKIP	United Kingdom Independence Party
UN	United Nations
UNCIP	United Nations Commission on India and Pakistan
UNRWA	United Nations Relief and Works Agency for Palestine Refugees in the Near East
WTO	World Trade Organization

PROLOGUE

What is in this book, and what will I get out of reading it?

The purposes of this book are:

- to provide a framework for interpreting and understanding current events;
- to introduce an academic understanding of geopolitics;
- to relate "real-world" events and development to theoretical perspectives;
- to place current events into a bigger picture;
- to understand geopolitics as an interaction between geopolitical agents and structures;
- to understand contemporary geography as an investigation of how politics and space are inseparable.

The book is an outgrowth of a class I taught for seven years at Penn State University and have continued to teach at the University of Illinois. The classes and the book aim to help people gain a greater understanding of current events by placing them in a bigger picture. I try to achieve this goal by introducing one overarching framework and a number of different theoretical perspectives. The overarching framework is one of structure and agency, or, put simply, that people and groups of people try to achieve particular goals within particular situations. These situations can, to some degree, help the people and groups (or agents) achieve their goals and/or frustrate them or limit their options. Using a geographic perspective, I define structures as both the situation within particular places and also how global politics has an impact upon actions taken within countries and places.

Though this book may be labeled a "textbook" I have tried to write it in a way that will be useful for both students and non-students who are seeking a way to gain a greater understanding of current events and how they relate to each other. With that goal in mind, I have attempted to balance the introduction of new concepts with exemplifications that have contemporary meaning, brief case studies of what are likely to be persistent conflicts, and guidelines and questions for you to pursue your own inquiries.

My goal has been to provide an *introduction* to geopolitics. The relatively recent upsurge of academic interest in geopolitics has produced many excellent books designed for graduate students, academics, professionals, and advanced undergraduates. This book provides an examination of geopolitics for students at the beginning of their under-graduate degrees, and so provides a stepping-stone to more theoretical approaches. My aim is to ease and promote access to the more advanced books so that more students and professionals are equipped with a critical understanding of geopolitics.

I feel lucky and privileged to have had the opportunity to teach and write on these topics. I hope that you find my way of looking at current events or geopolitics helpful. I wish you the best in your pursuit of greater understanding of our troubled world and hope that this book is of some help to you.

After reading this book you will:

- understand how politics create spaces and geographies;
- understand how politics are shaped by geographic circumstances;
- have a basic understanding of key persistent global conflicts;
- be able to use theoretical perspectives to develop a deeper and critical understanding of current events than can be gained by simply reading the newspaper or watching TV news.

A FRAMEWORK FOR UNDERSTANDING GEOPOLITICS

In this chapter we will:

- situate geopolitics within the discipline of human geography;
- define the key concepts of place and scale;
- introduce the concepts of structure and agency;
- show how place, scale, structure and agency will be used to understand geopolitics;
- define geopolitics;
- introduce a brief history of geopolitics;
- consider what is "power";
- provide examples of these concepts;
- use our own experiences and knowledge to understand and investigate these concepts.

Geopolitics: a component of human geography

Geopolitics is a component of human geography. To understand geopolitics we must first understand what is human geography. This is easier said than done, precisely because geography is a diverse and contested discipline—in fact, the easiest, and increasingly accurate, definition is that human geography is what human geographers do: accurate, but not very helpful.

Geography is a peculiar discipline in that it does not lay intellectual claim to any particular subject matter. Political scientists study politics, sociologists study society, etc. However, a university geography department is likely to house an eclectic bunch of academics studying anything from glaciers and global climate change, to globalization, urbanization, or identity politics. The shared trait is the *perspective* used to analyze the topic, and not the topic itself. Geographers examine the world through a geographic or spatial perspective, offering new insight in related disciplines. For example, a political geographer may study elections or wars (as would a political scientist or scholar of international relations) but argue that full understanding is only available from a geographic perspective.

So what is a geographic perspective? In the modern history of the discipline, dominant views of what the particular perspective should be have come and gone. In the middle of the twentieth century there was an emphasis upon geography as a description and synthesis of the physical and social aspects of a region. Later, many geographers adopted a mathematical understanding of spatial relationships, such as the geographic location of cities and their interaction. Today, human geography is not dominated by one particular vision but many theoretical perspectives, from neo-classical economics through Marxism, feminism, and into post-colonialism, and different forms of postmodernism. Furthermore, it would also be hard to think of a social or physical issue that is *not* being addressed by contemporary geography (see Hubbard *et al.*, 2002 and Johnston and Sidaway, 2004 to understand the history of geography and the variety of its current content).

Geography and places

But still, some guiding definitions are available to introduce readers to the discipline, if necessary, and also the content and purpose of geopolitics. Human geography may be defined as: "Systematic study of what makes places unique and the connections and interactions between places" (Knox and Marston, 1998, p. 3). In this definition, human geographers are seen to focus upon the study of particular neighbourhoods, towns, cities, or countries (the meaning of place being broad here). In other words, geographers are viewed as people who study the specifics of the world, not just where Pyongyang is but what are its characteristics. "Characteristics" may include weather patterns, physical setting, the shape of a city, the pattern of housing, or the transport system. Political geographers are especially interested, among other things, in topics such as how the city of Pyongyang, for example, is organized to allow for political control in a totalitarian country.

However, places (whether neighborhoods or countries) are not viewed as isolated units that can only be understood through what happens within them. The first definition also highlights the need to understand places in relation to the rest of the world. Are they magnets of in-migration or sources of out-migration? Are investors of global capital seeking to put their money in a particular place, or are jobs being relocated to other parts of the world? Is a place a site for drug production, such as areas of Afghanistan, or the venue for illegal drug use, such as suburban areas of the United States or Europe? Understanding a place requires analyzing how its uniqueness is produced through a combination of physical, social, economic, and political attributes—and how those attributes are partially a product of connections to other places, near and far.

Geography and spaces

A further and complementary definition of human geography is: "The study of the spatial organization of human activity" (Knox and Marston, 1998, p. 2). In this definition space is emphasized rather than place. The term space is more abstract than place. It gives greater weight to functional issues such as the control of territory, an inventory of objects (towns or nuclear power stations for example) within particular areas, or hierarchies and distances between objects. For example, a spatial analysis of drug production and consumption would concentrate on quantifying and mapping the flows of the drug trade, while an emphasis of place would integrate many influences to understand why drugs are grown in some places and consumed in others.

The economic, political, and social relationships that we enjoy and suffer are mediated by different roles for different spaces. Two banal examples: if you are going to throw a huge and rowdy party, don't do it in the library; as a student, when entering a university lecture hall sit in one of the rows of seats rather than standing behind the lecturer's podium. The banality of these examples only goes to show that our understanding of how society is spatially organized is so embedded within our perceptions that we act within sub-conscious geographical imaginations. In addition, these two examples also show that the spatial organization of a society reflects its politics, or relationships of power. Standing behind the lecturer's podium would be more than an invasion of her "personal space" but a challenge to her authority: it would challenge the status quo of student-lecturer power relationships by disrupting the established spatial organization of the classroom.

Compare the maps of Africa in Figure 1.1. The maps display two spatial organizations of power relations. The large map illustrates the spaces of independent countries (or states) that were created after the decline of the colonial control imposed by European powers in the nineteenth century. External powers defined parts of Africa as "theirs," and so allowed them to subjugate the native populations for perceived economic benefit free of violent and costly competition with other European countries. These spaces were a product of two sets of power relations: the ability of European countries to dominate African nations and the relative power parity among the European countries. The map of countries is a different spatial organization of power in Africa, the post-colonial establishment of independent African countries. This new spatial organization of power reflects a relative decrease in the power of the European countries to dominate Africa, though a hierarchy of power remains. However, focusing on the spaces of independent countries across the continent obscures other power relations, especially those of gender, race, and class relations *within* the countries. As we shall see, the scale at which we make our observations highlights some political relations and obscures others. The three smaller maps depict the struggle of Africans to end white-rule of their countries after the end of the colonial period. It shows the racialized spaces of political control, the areas of Africa in which the descendants of European settlers were able to maintain control, and how these spaces of white control have shrunk to nothing now given the end of apartheid in South Africa.

Places and politics

First, let's focus more closely on defining what we mean by place. From our earlier definition of human geography we know that places are unique and interdependent. In addition, "places provide the settings of people's daily lives" (Knox and Marston, 1998, p. 3). In other words, people's daily experiences, whether it be dodging mortar rounds

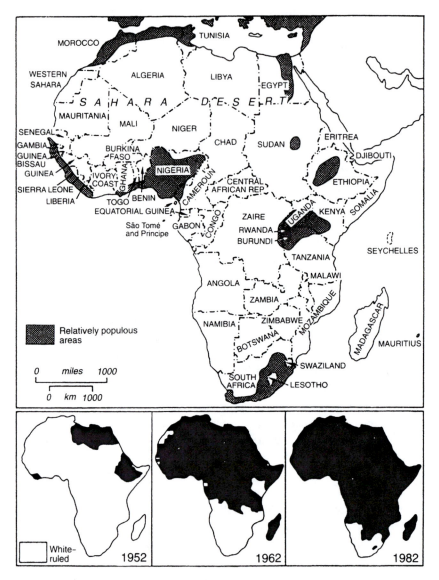

Figure 1.1 Africa: independent countries and the decline of white-rule.

in Baghdad or enjoying the wealthy trappings of up-scale housing in the gentrified London docklands, are a reflection of where they live. Life-chances are still very much determined by where one is born and grows up. Table 1.1 shows infant mortality rates across the globe, but also how this varies within countries, the United States is used as an example. The life expectancy in the state of Delaware, for example, is similar to the national average in Chile. What we may do, what we are aware of, what we think and "know" are a function of where we live. Places are the sites of employment, education, and conversation. Since places are unique they will produce a mosaic of experiences and understandings.

Table 1.1 The geography of infant mortality rates (per 1,000 live births)

Variation across countries (2000)		Variation within the United States (2001)	
United States	6.9	New Hampshire	3.8
Austria	4.8	Utah	4.8
Greece	5.0	California	5.4
Romania	19.0	Arizona	6.9
Chile	11.0	Georgia	8.6
Colombia	20.0	Alabama	9.4
Singapore	2.9	Mississippi	10.5
China	32.0	District of Columbia	10.6
Vietnam	23.0	Delaware	10.7
Pakistan	81.0		
Egypt	40.0		

Source: US Census Bureau; World Bank, World Development Indicators database, 2005.

To better understand how geographers think about place, we will use two different authors. First, John Agnew's (1987) definition suggests that places are the combination of three related aspects: location, locale, and sense of place.

Location is the role a place plays in the world, or its function. The key industries and sources of employment within a place are a good measure of location—whether it is a steel mill, coal mine, military base, or tourist resort. Of course, these are simplistic examples, and usually places will be a combination of different functions—perhaps complementing each other or existing together uneasily. Dover, England is an example where its function as a ferry port has promoted another function, the point of entry for refugees.

Activity

Stop reading for a minute and write down four or five features of your hometown that make it distinctive. We will be referring back to these features and developing them as we go through this chapter.

My example: I grew up near Dover, England. It is a major ferry port connecting the British Isles to the European Continent. Surrounding the town was a scattering of coal mines that were closed down after the miner's strike of 1984. The landscape of Dover is dominated by Dover castle situated on the cliffs, the castle keep dates from Norman times, and within the grounds are the ruins of a Roman lighthouse. The opening of the Channel Tunnel has threatened the profitability of the cross-channel ferries. In the past few years the town has experienced the presence of refugees from eastern and southeastern Europe. Though I am emotionally connected to Dover, I do find it a bit drab.

Locale refers to the institutions that organize activity, politics, and identity in a place. People operate as parts of groups: families, schools, workplaces, communities of worship, labor unions, political parties, militias, parent-teacher organizations, sports clubs, etc. In combination, these institutions form the social life of a particular place. The geography of the battle for control of post-Hussein Iraq was one of competing institutions with place-specific concentrations of strength. The city of Najaf and the area of Baghdad known as Sadr City housed insurgents mobilized and sustained through particular Shi'ite institutions. On the other hand, insurgents in Ramallah were maintained by Sunni religious institutions. Religious allegiance, maintained by certain Islamic institutions, facilitated the mobilization of insurgent groups in these places. This geography of the political and religious institutions in Iraq was a major reason for the difficulties in negotiating a new constitution for Iraq in 2005. The Kurdish and Shia regions of Iraq wanted more regional autonomy, while the minority Sunni Arabs feared that such a political organization of space would leave them marginalized.

On a very different note, a key element of the uniqueness of the wealthy town of Henley on Thames in Oxfordshire, England is the Henley Rowing Club and its sponsorship of an annual regatta that attracts England's social elite. In Dover, the presence of refugees has become an important part of local politics, provoking concerns that activity by far-right political groups and parties may increase.

The third aspect of place is *sense of place*. People's identity is a function of membership in a number of collective identities; gender, race, social class, profession, nationality, and, last but not least, place. Sense of place is a collective identity tied to a particular place, perhaps best thought of as the unique "character" of a place. People are guided in their actions by particular identities that say who they are and what they can and cannot, should and should not, do. Belonging to a particular ethnic group socializes people into particular expectations and life-chances. Part of one's sense of "belonging" is attachment to place, which can translate into visions of what a place should be "like": notably who "belongs" and who doesn't. A harmless example is the urban myth underlying the self-proclaimed moniker of the "Dover Sharks." The name is dubiously derived from stories of piracy from centuries ago when the locals would set lights to confuse the channel shipping and induce wrecks. They would then wait for and loot the cargo that washed ashore. The name "Dover Sharks" is claimed to derive from the habit of freeing rings from the bloated corpses by biting off their fingers! More disturbing is the way that Dover residents have reacted to the refugee "outsiders" as undeserving recipients of local government services that should only be available to residents of Dover. To relate place identify to contemporary conflict, the quote in Box 1.1, is an observation of a Palestinian man whose political beliefs are clearly tied to his attachment to place.

Using a sense of place to construct an identity politics of insiders and outsiders (Cresswell, 1996) is evident in the language of hate politics in the US when those deemed "not to belong" enter places with dominant or established "identities" and "traditions." Such politics are frequently racial, but homosexuals are regularly targeted too. Gentrified neighborhoods that are known to have large gay populations are often associated with anti-gay hate crimes that are spurred not solely by homophobia, but also by indignation over how the place has changed from an understanding of its "traditional" form. For

Box 1.1 Place and Palestinian identity

The man who entered the room was visibly distraught. Wasting no time on pleasantries, he threw himself down in a chair and announced that the soldiers had gone berserk. This was in early 1994, before the Israeli pull-back. Just before midnight, the man said, twenty or thirty people in the al-Boureij refugee camp had been forced out of their homes. Some of them hadn't even had a chance to put on their shoes; others complained that the soldiers had kicked and hit them. They were led to the UNRWA school (United Nations Relief and Works Agency for Palestine Refugees in the Near East) and ordered to pick up some garbage and rocks that had been strewn in the yard. Furious, the man said that someone was made to write slogans in Arabic on the wall. "Life is like a cucumber," was the worst of them. "One day in your hand, the next day up your ass."

During the four years of the intifada, Kafarna [the man who entered the room] had witnessed countless violent clashes and far greater indignities than those he recounted that day, but he never managed to come to terms with any of it; he was simply unable to swallow the insult.

(Hass, 2000, p. 31)

In this quote, note not only the evocation of the Gaza Strip as a place, but also how Kafarna's individual identity and politics are inseparable from his place-specific experiences.

example, particular visions of place are evident in anti-gay hate crimes in the Victorian Village neighborhood of Columbus, Ohio:

When you have somebody who, say, has a rainbow flag on their house and then you drive by and you see that it's been torched, it's not just a crime against the people living in that house. That is sending a message to the entire community the same way that having somebody put a cross on somebody else's front lawn sends a message to the entire community.

(Executive Director, Buckeye Region Anti-Violence Organization, May 29, 2001, quoted in Sumartojo, 2004, p. 100)

In another example, the murder of Dutch filmmaker Theo van Gogh by Islamic radicals in the Netherlands in November 2004 sparked a debate about the political future of a country proud of its record of tolerance. One Amsterdam newspaper, the *Algemeen Dagblad*, claimed that anti-Muslim graffiti suddenly appeared everywhere, and attacks on mosques and other Islamic buildings erupted. However, the Amsterdam newspaper *Trouw* argued that such reports were overblown and reported the Prime Minister Jan-Peter Balkenende as saying:

> We must not allow ourselves to be swept away in a maelstrom of violence
> . . . Free expression of opinion, freedom of religion, and other basic rights are
> the foundation stones of our state . . . all moderates will join together in the
> fight against the common enemy: extremism.
>
> (All quotes from *The Week*, November 26, 2004, p. 15)

As a general conclusion, the function of a place, which social groups have control of the institutions within a place, and the identity of a place, are contested. Racialist and homophobic groups are extreme examples of the politics of place.

Activity

Refer back to the features of your hometown you identified previously.

• Classify these features using Agnew's three aspects of place. If one of the aspects is not included think of a feature of your hometown that would fit.

• Can you find examples in your local and national newspapers of how the location, locale, and sense of place of your hometown have been or still are contested?

An alternative view of society emphasizes cosmopolitanism—attachment to no particular place. Globalization is seen by some to have created a class of "global citizens" who travel across the globe on business and political trips, or even for leisure. Focusing upon this relatively small group of people should not detract from the fact that it is the socially privileged (in terms of wealth, race, and gender) who have the status and ability to travel easily from place to place and feel at "home" wherever they are (Massey, 1994, p. 149).

Perhaps more worthy of our attention is the role of diasporas—networks of migrants who establish connections between places across the globe. A good example is the Chinese community in Vancouver, Canada that has facilitated massive amounts of investment by Chinese capitalists (large and small) in the real estate economy of British Columbia. Diasporas illustrate how a person can be attached to a number of places, though perhaps this geography may mean they are not completely "at home" anywhere.

The second author we will discuss is Doreen Massey. Her definition of place complements John Agnew's. For Massey, places are networks of social relations

> which have over time been constructed, laid down, interacted with one another,
> decayed and renewed. Some of these relations will be, as it were, contained
> within the place; others will stretch beyond it, tying any particular locality into
> wider relations and processes in which other places are implicated too.
>
> (Massey, 1994, p. 120)

Massey's definition gives us three extra points to consider about places. First, they are the products of human activity, or, in social science parlance, they are "socially constructed." The functions of a place, the institutions within it, and its character all stem from what people do. In 2004, Najaf as a place of insurgency against US and Iraqi security forces was a product of individual actions, the groups they formed, and the construction of a rebel or resistance identity within national and religious histories. When referring to "social relations," Massey identifies social hierarchies formed within the workplace, between racial and religious groups, and also the pervasive influence of gender upon normative expectations.

Second, places are dynamic or they change over time. What people, do, want, and think changes over time and such aspirations are translated into projects that make and remake places. Najaf is different now from the time I wrote these words. Insurgency, defeat, and negotiation have combined to change Najaf's role—maintaining its important religious function and hopefully a peaceful city within a peaceful Iraq. In another example, the landscape of contemporary Moscow is made up of layers from the Soviet past and its celebration of Communism and contemporary signs of consumer capitalism (see Figure 1.2).

Third, and related to our first definition of human geography, places can only be understood fully through their interactions with other places. Najaf's plight is understood in terms of resisting centralized control from Baghdad and its refusal to see soldiers sent from the US as a legitimate police force in a new Iraq. Also, it was argued that insurgency in the town was facilitated by military and political support from across the nearby Iranian border.

Figure 1.2 Lenin Statue, Moscow.

Activity

- How have the location, locale, and sense of place of your hometown changed over time?
- How is your hometown connected to other places?
- How does your hometown's past history influence its present and future?

Massey's emphasis upon the dynamism of place and Agnew's recognition of institutional politics and sense of place illustrate the central role contest or conflict plays in defining places. Let's go back to our earlier banal examples of the interaction between space and politics. Partying in the library would be an act that challenged the norms and rules of a particular place: a political act to change the function, meaning, and ambience of the library.

In a more significant example, Okinawa, Japan has been dominated by US military bases since the end of World War II. In Agnew's terminology, Okinawa's location is defined by its geostrategic military role for the US. An article in the *New York Times* of September 13, 2004 highlighted the aftermath of a military helicopter crash in which the Japanese police were not allowed to investigate the crash-site. Through Agnew's lens, different institutions were in a jurisdictional contest over the territory of Okinawa; who is in control: the US military or the Japanese police? Such contestation led to local protest, and expressions of self-identity reflected in a local high school teacher's sentiments: "At that time I felt Okinawa is really occupied by the US, that it is not part of Japan" (p. A4).

The role of Okinawa, the power of the local police in relation to the US military, and the identity of the place are contested, especially inflamed by incidents such as the helicopter crash. Places contain many different institutions and collective identity is usually multi-dimensional. So, places are sites of multiple conflicts. In the case of Okinawa, the conflict over the US presence is connected to the situation of the island within Japan. The *New York Times* article went on to quote the female teacher claiming "Tokyo doesn't care . . . I feel a gap between Tokyo and here" (p. A4). In other words, the contestation of what Okinawa is and will become is a combination of a regional identity rejecting the authority of the Japanese government and an assertion of local authority over the American presence.

What other contests are likely in play here but not mentioned in this article? Some hints include the gender of the teacher and the mention in the article of demonstrations a decade earlier after the rape of a 12-year-old schoolgirl by three American servicemen.

The protests in Okinawa illustrate many points we have covered: Agnew's three aspects of place should be understood as connected rather than separate entities; the nature of a place is a function of its connections to the outside world; places are contested; and the contestation produces dynamism—places change. The Okinawa example also introduces us to another important geographic concept, scale. It is impossible to interpret the actions of the Okinawan high school teacher without taking into consideration the island, its position within Japan, and the US's global military presence. It is to geographic scale that we now turn.

Activity

Now is the time for you to begin using the concepts introduced in the text to interpret media reports of current affairs. Look through current newspaper reports and find one that addresses the politics of a particular place.

- What components of the definitions of place provided by Agnew and Massey can you find in the article?
- What contests are in play?
- How do collective identities such as race, gender, age, class, and nationality interact with the concepts identified by Agnew and Massey?

The politics of scale

The actions of individuals and groups of individuals range in their geographic scope or reach. It is this scope or reach that is known as geographic scale. Place is one geographic scale, defined as the setting of our everyday lives. But place is just one scale in a hierarchy that stretches from the individual to the global (Taylor and Flint, 2000, pp. 40–6). (Perhaps even these boundaries are too narrow; genetic material and outer-space could arguably be seen as the geographical limits upon human behavior.) As a simple example, let's talk about economics. Well, do you mean ones own personal financial situation, the "family fortune" or lack of it, the local economy, national economic growth or recession, the economic health of the European Union or the North American Free Trade Agreement (NAFTA) region, or the global economy? Each of these scales represents a different level of economic activity, or transactions that define local economic health or the trade and investment that spans the globe.

But in a hierarchy scales are "nested" or connected (Herb and Kaplan, 1999; Herod and Wright, 2002). To illustrate the point, if all businesses were thriving, then all local economies would be booming, every national economy growing, and the global economy healthy. But, of course this is never the case; the viability of a business is partially defined by the opportunities within its scope. The family-owned hardware store or photocopying franchise is dependent upon enough wealthy customers nearby. A global company, such as Honda or Nike negotiates the differential opportunities for sales in different countries. In turn, the relative prosperity of individuals is related to the economic health of the businesses they work in and those businesses' national and global markets.

Political acts also negotiate scale. Protest can be enacted at the individual scale, by breaking laws seen by the individual as unjust or by wearing clothes or tattoos that make a political statement. An action such as not singing a national anthem when it is demanded or expected is another example of political action at the individual scale. But protest can also involve vigils outside of, say, abortion clinics or protests at animal hunts or laboratories conducting tests on animals. These "localized" acts require individual commitment and are also often motivated by national campaigns aimed at influencing the national legislative process. Increasingly, protest politics do not stop at the national scale; abortion politics, for example, are a component of discussions over the form of US foreign aid as well as a component of the missionary activity of many churches.

The examples show that geographic scales, like places, are socially constructed or made by human activity. We wear certain clothes and act in certain ways to create our own personae. Political parties and social movements are formed and maintained by individual activity, whether it be the highly public and visible speeches of the leader or the "bake sales" and envelope-stuffing activities of committed members. As the scope of the geographic scale increases it is harder to envision how they are socially constructed; but the everyday practices of paying taxes, maintaining national armed forces, politicking for the "national interest," and cheering on national teams in the Olympic Games or World Cup ensure the functional expression of a country, and the sense of national identity. In the workplace, we act to produce and consume products that are the outcome of economic activities from across the globe, unconsciously we reproduce the tea-leaf pickers in Sri Lanka, as well as the brokers who trade the picked leaves, the bankers who finance the plantations, and the advertisers who suggest the merits of having a "cuppa" on a regular basis. Though scales are made by human activity, the larger their scope, the less aware we are of the implications of our actions, and their importance in sustaining operations at that scale.

Participating in elections is another example of how scales are constructed by people. By choosing to vote or not to vote, an individual chooses to become involved in a particular way with the political system, either validating it or not. The aggregate of individual votes in a particular constituency creates a political jurisdiction as a particular political locality; either a "safe" or "contested" seat. In addition, the outcome of individual votes creates a national political system; either maintaining established democratic practices or forcing a change in the political system. For example, in 2004 the sustained demonstration of Ukrainian citizens, as well as the actions of government ministers and army officers who prevented a violent suppression of the actions, overcame electoral fraud to elect a new president. Finally, the example of elections shows that scales, just like places, are contested. The individual may well compromise their own beliefs by voting for a party; for example voting for a party because of their views on membership in the European Union despite being uncomfortable about, say, social or educational policies. Furthermore, the constituency scale and national scale are the very product of competing political parties.

The contested nature of scales requires us to think more closely about how scales are made by political actions, and with that in mind we will discuss the concept of geopolitical agents, a key concept for this book. But before we do that, we will give greater attention to the term geopolitics and briefly discuss the history of geopolitical theory.

Activity

Consider how your actions occur within a hierarchy of scales.

- If you are a member of a political organization (a party or pressure group) or a church, a member of the military services, or the employee of a business think about how you are connected to a small group, which is part of a bigger organization.

- Think about the influences upon you and the smaller group that stem from national and global events.

- Also, think about how your actions work up this hierarchy of scales to construct the organization, as well as the national and global scales.

What is geopolitics?

Geopolitics is a word that conjures up images. In one sense, the word provokes ideas of war, empire, and diplomacy: geopolitics is the practice of states controlling and competing for territory. There is another sense by which I mean geopolitics creates images: geopolitics, in theory, language, and practice, classifies swathes of territory and masses of people. For instance, the Cold War, was a conflict over the control of territory that was provoked and justified through geographically based images of "the Iron Curtain" (see Box 1.2) and the "free world" and the "threat" of Communism from the perspective of Western governments and the "imperialism" of America from the Soviet Union's view (Figure 1.3).

So how should we define geopolitics, in the contemporary world and with the intent of offering a critical analysis? Our goals of understanding, analyzing, and being able to critique world politics require us to operate with more than one definition.

First, we must note the connection between geopolitics and statesmanship: the "practices and representations of territorial strategies" (Gilmartin and Kofman, 2004, p. 113). For now, we will take a limited perspective on this definition—and note how states or countries have competed for the control of territory and/or the resources within them. At the end of the nineteenth century, the European powers indulged in an unseemly struggle for colonial control over Africa, what is known as the "scramble for Africa." In a contemporary sense, the geopolitics of the "War on Terrorism" has produced alliances between states and the deployment of troops in Afghanistan, Iraq, and in bases across Central Asia. Inseparable from these "practices" of fighting in Iraq and Afghanistan, for example, is the role of representation: the fight against "evil," the spread of "democracy," etc.

Second, geopolitics is more than the competition over territory and the means of justifying such actions: geopolitics is a way of "seeing" the world. From a feminist perspective, geopolitics is a masculine practice, hence my use of the term states*man*ship in the previous paragraph. In the much quoted words of Donna Haraway (1998), the practices and representations of geopolitics have relied upon "a view from nowhere."

Box 1.2 Winston Churchill's "Iron Curtain" speech

This box contains excerpts of Winston Churchill's (who had been Prime Minister of Britain during World War II) famous Sinews of Peace speech, made in 1946, in which he identified the "iron curtain" that would divide Europe throughout the Cold War. Coming just after the allied victory in World War II, in which Britain, the United States, and the Soviet Union fought on the same side, this was a rhetorical watershed in the public's awareness of the Cold War, and the identification of the Soviets and Communism as a threat to peace. Read the excerpts and find phrases that refer to (1) the control of territory by particular countries, and (2) the rhetoric or language used to either justify such control or identify it as a threat.

A shadow has fallen upon the scenes so lately lighted by the Allied victory. Nobody knows what Soviet Russia and its Communist international organization intends to do in the immediate future, or what are the limits, if any, to their expansive and proselytizing tendencies. I have a strong admiration and regard for the valiant Russian people and for my wartime comrade, Marshall Stalin. There is deep sympathy and goodwill in Britain—and I doubt not here also—towards the peoples of all the Russias and a resolve to persevere through many differences and rebuffs in establishing lasting friendships. We understand the Russian need to be secure on her western frontiers by the removal of all possibility of German aggression. We welcome Russia to her rightful place among the leading nations of the world. We welcome her flag upon the seas. Above all, we welcome, or should welcome, constant, frequent and growing contacts between the Russian people and our own people on both sides of the Atlantic. It is my duty however, for I am sure you would wish me to state the facts as I see them to you. It is my duty to place before you certain facts about the present position in Europe.

From Stettin in the Baltic to Trieste in the Adriatic an *iron curtain* has descended across the Continent. Behind that line lie all the capitals of the ancient states of Central and Eastern Europe. Warsaw, Berlin, Prague, Vienna, Budapest, Belgrade, Bucharest and Sofia, all these famous cities and the populations around them lie in what I must call the Soviet sphere, and all are subject in one form or another, not only to Soviet influence but to a very high and, in some cases, increasing measure of control from Moscow. Athens alone—Greece with its immortal glories—is free to decide its future at an election under British, American and French observation. The Russian-dominated Polish government has been encouraged to make enormous and wrongful inroads upon Germany, and mass expulsions of millions of Germans on a scale grievous and undreamed-of are now taking place. The Communist parties, which were very small in all these Eastern States of Europe, have been raised to pre-eminence and power far beyond their numbers and are seeking everywhere to obtain totalitarian control. Police governments are prevailing in nearly every case, and so far, except in Czechoslovakia, there is no true democracy. . . .

The safety of the world, ladies and gentleman, requires a new unity in Europe, from which no nation should be permanently outcast. It is from the quarrels of the strong parent races in Europe that the world wars we have witnessed, or which occurred in former times, have sprung. Twice in our own lifetime we have seen the United States, against their wishes and their traditions, against arguments, the force of which it is impossible not to comprehend, twice we have seen them drawn by irresistible forces, into these wars in time to secure the victory of the good cause, but only after frightful slaughter and devastation have occurred. Twice the United States has had to send several millions of its young men across the Atlantic to find the war; but now war can find any nation, wherever it may dwell between dusk and dawn. Surely we should work with conscious purpose for a grand pacification of Europe, within the structure of the United Nations and in accordance with our Charter. That I feel opens a course of policy of very great importance.

Figure 1.3 The Iron Curtain.

As we will see soon, geopolitical theoreticians have made claims that they can view or understand the whole globe. In other words, they operate under the belief that the whole world is a "transparent space" that is "seeable" and "knowable" from the vantage point of the white, male, and higher class viewpoint of the theoretician (Staeheli and Kofman, 2004, p. 4 referring to Haraway (1998) and Rose (1997)). Geopolitical theoreticians classify the world into particular regions while also defining historical trends. The feminist critique rests on the idea that all knowledge is "situated" and, hence, "partial." The very fact that the classical geopoliticians were from privileged class, race, and gender backgrounds in Western countries meant that they had absorbed particular understandings of the world; they were unable to know the *whole* world. In stark contradiction, their policy prescriptions rested upon the assumption and arrogance of being able to see and know the whole world and the essence of its historical development.

A third understanding of geopolitics results from the identification of "situated knowledge": geopolitics is not just a matter of countries competing against countries, there are many "situations" or, in other words, the competition for territory is broader than state practices. Geopolitics has come to be understood as much more than war and the building of empire. It can also include racial conflicts within cities, the restrictions upon the movement of women in certain neighborhoods and at certain times because of patriarchal laws and/or fear of attack, and diplomacy over greenhouse gas emissions, as a few examples. In other words, geopolitics is not the preserve of states; individuals, protest movements, non-governmental organizations (NGOs) such as Greenpeace and Amnesty International, terrorists, and private companies are all engaged (and always have been) in the control of territory and so have struggled to represent it in certain ways. Following Gilmartin and Kofman (2004), geopolitics is the multiple practices and multiple representations of a wide variety of territories.

Following on from the third definition, we identify our fourth and final meaning of the word. Geopolitics has come to include *critical geopolitics* (Ó Tuathail, 1996), which is the practice of identifying the power relationships within geopolitical statements; what assumptions underlie phrases such as "the spread of free markets" or the "diffusion of democracy", for example. What are the consequences of such practices and representations? Who gains what, and who suffers? In other words, the phrases commonly used to justify state practices are put under critical scrutiny to see how they try to restrict our view of the world and promote a limited number of policies. By promoting interpretations of world events that are counter to dominant government and media representations, critical geopolitics aims to encourage anti-geopolitics; practices by individuals, groups of citizens, indigenous peoples, etc. to resist the control and classifications imposed by states and other powerful institutions such as the World Bank.

Contemporary geopolitics identifies the sources, practices, and representations that allow for the control of territory and the extraction of resources. States still practice statesmanship. In that sense we are still offered "all seeing" interpretations of the world by political leaders and opinion makers. But their "situated knowledge" has been increasingly challenged by others in "situations" different from the clubs and meeting rooms of politicians and business leaders. As a result, geopolitical knowledge is seen as part of the struggle as marginalized people in different situations aim to resist the domination of the views of the powerful. Feminist geopolitics has invoked the need for a "populated" geopolitics, one that identifies the complexity of the world, and the particular situations of people across the world, as opposed to the simplistic models of classic geopolitics and their simple explanations (Gilmartin and Kofman, 2004, p. 115).

Geopolitics is "up for grabs" as the type and goals of territorial conflicts have become increasingly broader. But before we move on, back to our roots, who were the classical geopoliticians and what did they do?

A brief history of geopolitics

As we have noted, geopolitical knowledge is "situated knowledge." Though this observation has been used to claim the relevance of the perspectives and actions of marginalized groups, it may still be used to consider the thoughts of the theoreticians whose concern was geopolitical states*man*ship. In other words, geopolitical theoreticians constructed their frameworks within particular political contexts and within particular academic debates that were influential at the time, the latter sometimes called paradigms.

Geopolitics, as thought and practice, is linked to the establishment of states and nation-states as the dominant political institutions. Especially, geopolitics is connected to the end of the nineteenth century—a period of increasing competition between the most powerful states—and it is the theories generated at this time that we will label "classic geopolitics" (Table 1.2). Geopolitics was initially understood as the realm of inter-state conflict, with the quiet assumption that the only states being discussed were the powerful Western countries. In other words, there was a theoretical attempt to separate geopolitics from imperialism, the dominance of powerful countries over weaker states.

Sir Halford Mackinder (1861–1947) is, perhaps, the most well-known and influential of the geopoliticians who emerged at the end of the nineteenth century. The kernel of his idea was used in justifying the nuclear policy of President Reagan, and academics and policymakers continue to discuss the merits of his "Heartland" theory. The political context from which Mackinder wrote was multi-layered. Internationally, he was concerned about the relative decline in Great Britain's power as it faced the challenge of Germany. Within Britain, his conservatism was appalled by the destruction of traditional agricultural and aristocratic lifestyles in the wake of industrialization, especially the rise of an organized working class that made claims for social change. His goal was

Table 1.2 Features of classic geopolitics

Privileged position of author	White, male, elite, and Western situated knowledge
Masculine perspective	"All seeing" and "all knowing"
Labeling/classification	Territories are given value and meaning
A call to "objective" theory or history	Universal "truths" used to justify foreign policy
Simplification	A catchphrase to foster public support
State-centric	Politics of territorial state sovereignty

to maintain both Britain's power and its landed gentry through a strong imperial bloc that could resist challengers while maintaining wealth and the aristocratic social structure.

Influenced by the work of Alfred Thayer Mahan (1840–1914), Mackinder saw global politics as a "closed system"—meaning that the actions of different countries were necessarily interconnected, and that the major axis of conflict was between land- and sea-powers. He defined the geography and history of land-power by defining, in 1904, the core of Eurasia as the Pivot Area, which in 1919 he renamed the "Heartland" (Figure 1.4). This area was called the Pivot Area because, in his Eurocentric gaze, the history of the world pivoted around the sequence of invasions out of this region into the surrounding areas that were more oriented to the sea. In the past, Mackinder believed sea-powers had maintained an advantage, but with the introduction of railways, he reasoned, the advantage had switched to land-powers; especially if one country could dominate and organize the inaccessible "Heartland" zone. Hence Mackinder's famous dictum, or, in contemporary language "bumper sticker":

> Who rules East Europe commands the Heartland.
> Who rules the Heartland commands the World Island.
> Who rules the World Island commands the World.

The World Island was Mackinder's term for the combined Eurasian and African landmasses.

Mackinder's twin goals were to maintain British global preeminence in the face of challenge from Germany, the country most likely to "rule" eastern Europe, and in the process, resist changes to British society. After initially discounting the role of the United States, in 1924 he proposed a Midland Ocean Alliance with the US to counter a possible alliance between Germany and the Soviet Union. Following his identification of the "Heartland", roughly representing the territorial core of the Soviet Union, plus his emphasis on alliance, Mackinder was the intellectual basis for Cold War strategists and proponents of the North Atlantic Treaty Organization (NATO).

Mackinder's contribution is also a good illustration of two prevalent features of "classic" geopolitics. First, he used a limited and dubious Western-centric "theory" of history to claim an objective, neutral, and informed intellectual basis for what is in fact a very biased or "situated" view with the aim of advocating and justifying the policy of one particular country. Plus he disseminated a catchy phrase or saying to influence policy. Second, Mackinder's career is one of many examples of the crossover between academic or "formal" geopolitics and state policy or "practical" geopolitics: he was a successful academic, founding the Oxford School of Geography in 1899 and serving as director of the London School of Economics between 1903 and 1908 and a member of Parliament.

Alfred Thayer Mahan (1840–1914) also walked in academic and policy circles. He rose to the rank of admiral in the US navy and was president, at different times, of both the Newport War College and the Naval War College. His two books *Influence of Seapower upon History* (1890) and *The Interest of America in Seapower* (1897) were important influences upon Presidents McKinley and Theodore Roosevelt, as well as the German Kaiser Wilhelm II. Mahan made a historical distinction between

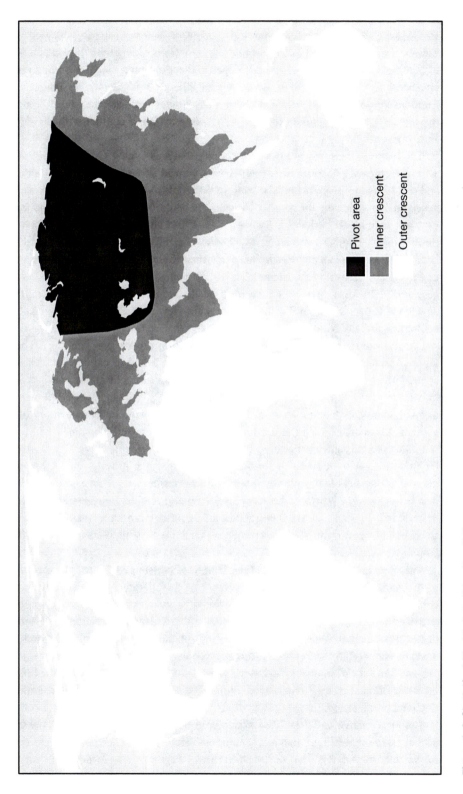

Figure 1.4 Sir Halford Mackinder's "Heartland" theory.

land- and sea-powers that was to influence geopolitical thinkers through the Cold War. He believed that great powers were those countries whose insularity, coupled with an easily defensible coastline, provided a secure base from which, with the aid of a network of land bases, sea-power could be developed and national and global power attained and enhanced. In addition, Mahan advocated an alliance with Britain to counterbalance Eurasian land-powers. His influence upon Mackinder is clear, but Mahan's goal was to increase US global influence and reach, while avoiding conflict with the dominant British navy.

The United States was not the only country that was eyeing Great Britain's supremacy. In Germany, politicians and intellectuals viewed Britain as an arrogant nation that had no "divine right" to its global power. In the words of Chancellor Bismarck, Germany deserved its "place in the sun." "German" geopolitics was defined by the work of two key individuals: Friedrich Ratzel (1844–1904) and Rudolf Kjellen (1864–1922). Similar to his English counterpart Mackinder, Ratzel was instrumental in establishing geography as an academic discipline. Furthermore, his *Politische Geographie* (1897) and his paper "Laws of the Spatial Growth of States" (1896) laid the foundations for *geopolitik*. However, it was the Swedish academic and parliamentarian Kjellen who developed Ratzel's idea and refined an organic view of the state. Following Ratzel's zoological notions, Kjellen propagated the idea that states were dynamic entities that "naturally" grew with greater strength. The engine for growth was "culture." The more vigorous and "advanced" the culture, the more right it had to expand its "domain" or control more territory. Just as a strong pack of wolves could claim hunting grounds of a neighboring but weaker pack, the organic theory of the state asserted that it was more efficient and "natural" for advanced cultures to expand into the territory of lesser cultures. Of course, given the existing idea that cultures were contained within countries or states this meant that borders were moveable or expandable. The catchphrase for these ideas was Ratzel's *Lebensraum*, or living space: meaning that "superior," in the eye of the beholder, cultures deserved more territory as they would use the land in a better way. In practice, the ideas of Ratzel and Kjellen were aimed at increasing the size of the German state eastwards to create a large state that the "advanced" German culture warranted, in their minds, at the expense of the Slavs who were deemed culturally inferior.

The German process illustrates a key feature of classic geopolitics: the classification of the earth and its peoples into a hierarchy that then justifies political actions such as empire, war, alliance, or neglect. This process of social classification operates in parallel with a regionalization of the world into good/bad, safe/dangerous, valuable/unimportant, peaceful/conflictual zones. Dubious "theories" of the history of the world and how it changes are used to "see" the dynamics of geopolitics as if from an objective position "above" the fray: Haraway's God's-eye view. Of course, we should note the influential positions of these geopoliticians. Geopolitical theorists are far from being neutral, objective, and uninterested.

Before we move on to the Cold War period, we should briefly return to the German school of geopolitics to make a couple of more points about classic geopolitics in general. As Adolf Hitler and the Nazi party began to rise to power in the 1920s, General Karl Haushofer (1869–1946) began to disseminate geopolitical ideas to the

German public through the means of a magazine/journal entitled *Zeitschrift für Geopolitik* (*Journal of Geopolitics*) and a weekly radio show. Haushofer was skillful in creating a geopolitical vision that unified two competing political camps in inter-war Germany: the landed aristocrats, who wanted to expand the borders of Germany eastwards toward Russia and the owners of new industries such as chemicals and engineering who desired the establishment of German colonies outside of Europe to gain access to raw materials and markets (Abraham, 1988). This idea came together in his definition of pan-regions (large multi-latitude regions that were dominated by a particular "core" power). In this scenario, the US dominated the Americas and Germany dominated Eurasia while Germany controlled Africa. Haushofer's vision allowed for both territorial growth and colonial acquisition for Germany, without initiating conflict with the US.

Haushofer blended a policy, and made the German public aware of foreign policy debates, that ran parallel with Hitler's surge in popularity and his vision of a "strong" Germany. However, Haushofer was not Hitler's "philosopher of Nazism" as *Life* magazine famously declared in 1939 (Ó Tuathail, 1996, p. 115). In fact, there was a significant difference between the views of Haushofer—with his emphasis on geographic or spatial relationships—and Hitler, whose racist view of the world shaped his geopolitical strategy. But, the point is that Haushofer did use Hitler's surge to power as a means of advancing his own career. Haushofer's tragic tale (he ultimately committed suicide following questioning by the US after the war regarding his role as a war criminal) has resonated throughout the community of political geographers ever since. Equating "geopolitics" with the Nazis tainted the sub-discipline of political geography and it practically disappeared as a field of academic inquiry immediately after World War II (see Box 1.3).

However, there is another lesson to take from Nazi geopolitics too—and that is how it continues to be portrayed by academics. Many recent studies have contextualized and examined the content of Nazi geopolitics in depth. Not to apologize for their connection to Hitler but to place the development of their theories within the contexts of global politics and the development of academic thought. The research shows there were indeed differences between their theories and Hitler's vision. Also, another outcome of this work is to show that Mackinder shared some of the academic baggage of the German geopoliticians. The predominance of biological analogies in social science at the end of the nineteenth century and beginning of the twentieth century meant that Mackinder and the German school were influenced by ideas that equated society with a dynamic organism. The key difference was that Mackinder was writing from, and for, a position of British naval strength, while the Germans were trying to challenge that power through continental alliances and conflicts with a wary and envious eye on British sea-power.

Post-World War II there existed an interesting irony: the vilification of "geopolitics" as a Nazi enterprise resulted in its virtual disappearance from the academic scene. On the other hand, as the United States began to develop its role as a post-war world power, it generated geopolitical strategic views that guided and justified its actions. Prior to World War II, Isaiah Bowman (1878–1956), one time president of the Association of American Geographers, offered a pragmatic approach to the US's global role, and was a key consultant to the government, most notably at the Treaty of Versailles negotiations at the end of World War I. Nicholas Spykman (1893–1943), a professor of International

Box 1.3 Environmental determinism and geopolitics

Geopolitics is the science of the conditioning of political processes by the earth. It is based on the broad foundation of geography, especially political geography, as the science of political space organisms and their structure. The essence of regions as comprehended from the geographical point of view provides the framework for geopolitics within which the course of political processes must proceed if they are to succeed in the long term. Though political leadership will occasionally reach beyond this frame, the earth dependency will always eventually exert its determining influence.

(Haushofer *et al.*, 1928, p. 27, quoted in O'Loughlin, 1994, pp. 112–13)

The quote from General Haushofer offers an example of the "geodeterminism" of classic geopolitics, or the way in which political actions are determined, as if inevitably, by geographic location or the environment. Such an approach can be used to justify foreign policy as it removes blame from decision-makers and places the onus on the geographic situation. In other words, if states are organisms then Germany's twentieth century conflicts with its neighbors are represented as the outcome of "natural laws" and not decisions made by its rulers.

Relations at Yale University, noted the US's rise to power and argued that it now needed to practice balance of power diplomacy, as the European powers had traditionally done. Similar to previous geopoliticians, Spykman offered a grandiose division of the world: the Old World consisting of the Eurasian continent, Africa, and Australia, and the New World of the Americas. The US dominated the latter sphere while the Old World, traditionally fragmented between powers, could, if united, challenge the United States. Spykman proposed an active, non-isolationist US foreign policy to construct and maintain a balance of power in the Old World in order to prevent a challenge to the United States. Spykman identified the "Rimland," following Mackinder's "inner crescent," as the key geopolitical arena. In contrast to the calls for greater global intervention, Major Alexander P. De Seversky (1894–1974) proposed a more isolationist and defensive stance. His theory is notable for its emphasis upon the polar regions as a new zone of conflict, using maps with a polar projection to show the geographical proximity of the US and Soviet Union, and the importance of air-power.

Increasingly, US geopolitical views took the form of government policy statements that, in the absence of academic endeavors, assumed the status of "theories," and hence gained an authority as if they were objective "truths." First came George Kennan's (1904–2005) call for containment, then the National Security Council's (NSC) call for a global conflict against Communism, in policy document NSC-68, supported by the dubious "domino theory." These geostrategic policy statements will be discussed in greater depth in Chapter 3. In the relative absence of academic engagement with the topic, geopolitical theories were constructed within policy circles, and, despite the global role of the US, a limited perspective remained. George Kennan, for example, is identified as a "man of the North [of the globe]" (Stephanson, 1989, p. 157) who identified the Third World as "a foreign space, wholly lacking in allure and best left to its own, no doubt, tragic fate" (p. 157). Kennan, in the tradition of his academic predecessors, was also eager to classify the world into regions with political meaning; defining a maritime trading world (the West) and a despotic xenophobic East.

Perhaps, in hindsight, the lack of policy-oriented geopolitical work in the academic world provided room for the critical understandings of geopolitics that now dominates the field. With the exception of Saul Cohen's (1963) attempt to provide an informed regionalization of the world to counter the blanket and ageographical claims of NSC-68, geographers were largely silent about the grand strategy of inter-state politics. However, with the publication of György Konrád's *Antipolitics* (1984), in accordance with other theoretical developments in social science thinking and public dissent over the nuclear policies of Ronald Reagan, geographers found a voice that produced the field of "critical geopolitics" as well as broader systemic theories about international politics. Both of these approaches, though very different in their content and theoretical frameworks, offered critical analysis of policy, rather than being a support for government policy.

Though it is hard to summarize the diversity of these critical approaches, there is one important commonality: the study of geopolitics is no longer state-centric. The world-systems approach contextualized the actions of particular countries or states within their historical and geographical settings. For example, the decisions made by a particular government are understood through the current situation in the world as a whole. It is this approach that guides most of the content of this book. Critical geopolitics "unpacked" the state by illustrating that it is impossible to separate "domestic" and "foreign" spheres, that non-state actors—such as multi-national companies, NGOs and a variety of protest groups and movements for the rights of indigenous peoples, minorities, women, and calling for fair trade, the protection of the environment, etc., play a key role in global politics.

The bottom line: geopolitics is no longer exclusively the preserve of a privileged male elite who used the authority of their academic position to frame policy for a particular country. Though these publications still exist (see the subsequent discussion of Samuel Huntington and Robert Kaplan, for example), most academics who say they study geopolitics are describing the situation of those who are marginalized, and advocating a change in their situation. Study of the state is often critical, but it just one component of a complicated world—rather than a political unit with the freedom to act as the theory suggests it should in a simplified and understandable world.

This brief history of geopolitics is intended to introduce you to the role and content of "classic" geopolitics and the growth of alternative geopolitical frameworks. Two words of caution. First, this history is Eurocentric. I urge the reader to use the *Dictionary of Geopolitics* (O'Loughlin, 1994) to see how thought in countries such as Japan and Brazil reflect and differ from those discussed earlier. Japan, for example, had its own debate about the merits of the German school of geopolitics, with the ideas of Ratzel and Kjellen being popular among Tokyo journalists but less so within academic circles.

Second, do not be fooled by the prevalence of "critical geopolitics" in the academy. Bookstores are continually replenished by volumes purporting to "know" everything about "Islam," "terrorists," and a variety of imminent or "coming wars." Some of these volumes are quite academic, and others more popular. They all share the arrogance of claiming to be able to predict the future and, hence, are assured about what policies should be adopted. "Classic geopolitics" lives, but now it must contend with an increasingly vigorous and confident "critical geopolitics." In other words, geopolitics is itself a venue and practice of politics.

Geography and geopolitics

Classic geopolitical theories are examples of "situated knowledge" that construct images of the world in order to advocate particular foreign policies. The "situation" of the knowledge is both social and geographical. All the classic theorists were white Eurocentric males with conservative outlooks and a degree of social privilege. But their "situation" can be analyzed through Agnew's geographic framework of location, locale, and sense of place. With regard to location, the relative economic strength of Britain, Germany, and the US drove the theorists' respective perceived foreign policy needs. The institutional settings of universities, government, and policy circles nurtured and disseminated the knowledge the theorists created. For example, both Mackinder's Eurocentrism and Kennan's derisive views of the Third World were generated through their socialization in particular family, social, educational, and professional settings that, in combination, made up a geographic locale. In sum, the classic geopoliticians carried a definite sense of place regarding their own country and other parts of the world, which was instrumental in formulating their geopolitical outlooks.

The theorists' classification of the globe into particular regions also reflected Agnew's framework. The strategic importance of a country or region was evaluated in terms of its location, both resource potential and strategic role. Despotism, colonial administration, and "free institutions" were the types of locale attributed to countries to define policies. Finally, in order to justify the policies, a sense of place had to be disseminated to the public, both the "goodness" and morality of one's own country, but also the threat and depravity of other countries. In other words, the classic geopolitical theorists constructed geographical images of the world (or maps of locations and locales) within their own place-specific settings. Our job now is to provide a framework for seeing how geopolitical actions are "situated" within the dynamics of the world.

Geopolitical agents: making and doing geopolitics

Up to now, I have referred to the actions of individuals and "groups of individuals." It is time to tighten up the language and answer the question, who or what conducts geopolitics? In social science parlance, we will identify geopolitical agents. By agency, we simply mean the act of trying to achieve a particular goal. A university student is an agent; their agency is aimed at completing their degree. A political party is an agent;

their agency is aimed at seeking power. A separatist movement is an agent; their agency is targeted toward achieving political independence. A country may also be seen as an agent; their agency is seen in their trade negotiations, for example.

In the nineteenth century, and throughout most of the twentieth century, geopolitics was viewed as the preserve of the state (or county) and statesmen, with the gender referent being important (Parker, 1985; Agnew, 2002, pp. 51–84). Geopolitics was the study, some claimed science, of explaining and predicting the strategic behavior of states. States were the exclusive agents of geopolitics. This is the period of "classic geopolitics" we discussed earlier. But, the contemporary understanding of geopolitics is much different; indeed one set of definitions would classify all politics as geopolitics, in a broad understanding that no conflict is separate from its spatial setting.

Hence we can talk of corporations involved in the geopolitics of resource extraction as they negotiate with governments for mineral rights and maintain security areas within sovereign countries, or the geopolitics of NGOs seeking refugee rights, or the geopolitics of nationalism, as a separatist group uses electoral politics and/or terrorism to push for an independent nation-state, for example. A provisional list of geopolitical agents could include: individuals, households, protest groups, countries, corporations, NGOs, political parties, rebel groups, and organized labor, though this list is far from complete. Similar to our discussion of geographic scale, it follows that these agents are not separate but entwined: an individual is a member of a household, citizen of a particular country and may be affiliated with a number of political organizations, as well as being employed within a firm. Thus, not only does an individual act out a number of geopolitics, the geopolitics may be competing.

Geopolitical agents work toward their goals, but their chances of success and the form of their strategy is partially dependent upon their context. They do not have freedom of choice, but they do have choices. They also do not act within a geopolitical vacuum; they make calculations based upon other agents.

Let us look at three examples. Iran's decisions regarding the pursuit of nuclear weaponry are made in a calculation of the power of other countries, two of which are also nuclear powers, Israel and the United States. Tony Blair's decision to provide military support for the 2003 invasion of Iraq and the post-war policing duties was made after calculating the response of members of the government, parliament, the Labour party, and the British electorate. The formation of the United Kingdom Independence Party (UKIP) is understood within the context of the Conservative party's electoral weakness, and the issue of British sovereignty within the European Union.

In the first example, the geopolitical agent is identified as a nation-state or country (Iran), and its calculations involve awareness of other countries, or agents of the same geographic scale. The second example, examines how the actions of the leader of a nation-state (Prime Minister Blair) required recognition of actions, or future actions, of agents at lower geographic scales, the political party and individuals. The third example identifies the UKIP as a geopolitical agent seeing opportunities for advancement because of the dynamics of a competing party as well as the events and changes at the scale of the European Union, or multi-national geopolitics.

Geopolitical agents can be thought of as geographic scales. Moreover, the nested pattern of geographic scales allows us to think of geopolitical agents as consisting of other agents and acting "below" or within yet more geopolitical agents. Our next conceptual task in this chapter is to explore what we mean by the use of the words "consisting of" and "within" in the previous sentence. We will do so through the terms structure and agency.

Structure and agency: possibilities, constraints, and geopolitical choices

The ideas of structure and agency are part of an intellectual debate within social science that can get us into some very complex philosophy. My goal here is to provide enough material for you to interpret contemporary geopolitics, rather than negotiating the philosophical debate. Provided below are some key rules to initially aid our discussion:

- Agents cannot act freely, but they are able to make choices.
- Agents act within structures.
- Structures limit, or constrain, the possible actions of the agent.
- Structures also facilitate agents, in other words they provide opportunities for agents to attain their goals.
- An agent can also be a structure and vice versa.

See Johnston and Sidaway (2004, pp. 219–64) and Peet (1998, pp. 112–93) for more on the theory of structures and agents and structuration theory.

What is a structure? A structure is a set of rules (formal as in legally enforceable laws) and norms (culturally accepted practices) that partially determine what can and cannot, could and should not, be done. In this sense, structures are expressions of power as they define what is permissible and expected. Agents are those entities attempting to act. In other words, a woman homemaker may be viewed as an agent, and the patriarchal household a structure. In another view, the very same household can be seen as an agent negotiating the laws and culture of a country, which is interpreted as the structure. And to take this further, that self-same country may be seen as an agent operating within the structure of the international state system with its international laws and diplomatic customs.

Why is this theoretical framework useful? First, it shows that agents are given both opportunities to act but also constraints to their possible actions given the structures they operate within. For example, a labor union may be given the ability to strike given the laws of the country, but the same laws may prevent blockading roads and other forms of civil disobedience. Second, agents will be able to use, and be frustrated by, a number of structures simultaneously, given the multiplicity of spheres they operate within. The labor union must also use friendly political parties and combat those that are critical, too. Third, we can see that a particular structure is not monolithic but made up of a number of agents. For example, the union consists of individuals who must take into consideration the needs of their own household. Hence, strikes can crumble as some union members vote for a return to work as financial pressures mount. No structure can be seen to be monolithic. Fourth, by knowing that agents are simultaneously structures and vice versa, we can think of the opportunities of agents and the barriers they face, within a hierarchy of geographic scale.

Thinking of the structures within which agents are operating as a hierarchy of scales allows us to identify the key spaces that are being fought over. In other words, we can define both the politics and the geography and, hence, the geopolitics in question. The agency of insurgents in Fallujah, for example, illustrates the importance of the national space of Iraq as a structure that gives the opportunity for rebellion. The ability of the Iraqi government to constrain the agency of the insurgents is partially a function of the global geopolitical structure, especially the global presence of the United States.

Finally, it must be stressed that structures are the products of agents. An insurgent group and security forces are made by the actions of their members, and the actions of the insurgency group and the security forces play a role in making the national space what it is. However, in addition the relationship is recursive. Or in other words, the national situation structures, to some extent, the actions of the insurgency group while those actions construct contemporary Iraq.

Activity

Reconsider how you located yourself within a hierarchy of scales in the previous exercise.

- In what way are you prevented from doing certain things because of norms and rules established at higher scales?

- In what way do norms, rules, and capabilities at higher scales allow you to do what you want to do?

- Also, in what way do your actions construct the norms, rules, and capabilities found at the higher scales?

Geopolitics, power, and geography

Geopolitics uses components of human geography to examine the use and implications of power. Contesting the nature of places and their relationship to the rest of the world is a power struggle between different interests and groups. The spatial organization of society, the establishment and extent (both geographic and jurisdictional) of state sovereignty is a continuing geopolitical process. The political aspirations and projects of geopolitical agents are won and lost within a structure of geographic scales. The fortune of geopolitical agents is also a function of their component parts, which can also be seen as geographic scales.

Scale, place, and space are arenas, products, and goals of geopolitical activity, and each of those three concepts have many different manifestations. Already, we have seen that a wide variety of geopolitical agents can be identified. In sum, it can be seen that conflicts over places, spaces, and scales are pervasive and multifaceted. To keep this book focused and manageable particular forms of geopolitical conflicts and particular geographies will be emphasized. Though this is necessarily exclusive, I also encourage you to explore other forms of geopolitics.

Geopolitics, as the struggle over the control of spaces and places, focuses upon power, or the ability to achieve particular goals in the face of opposition or alternatives. In nineteenth and early twentieth century geopolitical practices, power was seen simply as the relative power of countries in foreign affairs. For example, in the early 1900s US naval strategist Alfred Thayer Mahan's categorization of power was based upon the size of a country, the racial "character" of its population, as well as its economic and military capacity. In the late twentieth century, as the geopolitical study of power became increasingly academic, scholars created numerous indices of power, which remained focused on country-specific capabilities of industrial strength, size and educational level of the population, as well as military might. Definitions of power were dominated by a focus on a country's ability to wage war with other countries.

However, recent discussions of power have become more sophisticated. Feminists emphasize that the focus on government capabilities ignores other forms of power, such as gender and racial relationships within and between countries that are, over time, assumed to be "normal" or of secondary importance to the male-dominated practices of foreign policy. Feminist insistence of the integral role of gender relations in geopolitics leads to connections between the competitive nature of power relations between countries and the way patriarchal relations within countries normalize a masculine and militarized conception of foreign policy (Enloe, 1983, 1990, 2004). Feminism forces us to think about the gender and racial make-up of geopolitical agents and structures, so promoting the study of geopolitics as the combination of multiple power relations (consider Figure 1.5). The result is that any understanding of a current event must come from a variety of perspectives and not just the calculations of male-dominated elites.

One of the other contributions of feminist analysis is the focus upon how power relations become taken for granted or viewed as "common sense." Power in this sense is not the ability or need to force others to do what you want, but make them follow your agenda willingly without considering alternatives. These ideas stem from the Italian Marxist Antonio Gramsci (1971) who noted how a ruling class in a country needs to exert force to control the working classes only rarely. On the whole, subordinated groups "follow" political goals that are of greater benefit to the more powerful; alternatives are seen as "radical" or "unrealistic," while the dominant ideology is seen as "unpolitical" or "natural." For example, in the arena of international economics, policies for "economic development" created by the rich and powerful countries are adopted by the poorest countries of the world under the label of "progress" despite the growing global inequality levels after decades of such policies. The Gramscian notion of power requires us to consider how geopolitical practices and ideas are disseminated and portrayed to wide audiences in order to justify them and make them appear "normal" while belittling alternative views. In other words, the representation of geopolitics is another manifestation of power (Ó Tuathail, 1996).

In this book, I will stay closely tied to the original and traditional subject matter of geopolitics, states (more commonly referred to as countries). The classic geopoliticians of the late nineteenth and twentieth centuries expressed confidence in knowing "how

Figure 1.5 Woman and child in Iraqi bombsite.

the world works" and used a historical-theoretical perspective to suggest or justify the foreign policy actions, mainly aggressive, of their own countries (Agnew, 2002). My goal is not to explain away the acts of any country as the inevitable consequence of a deterministic world history. Instead, countries are seen as key geopolitical agents, and their actions are understood through examining the competition between agents within a country, as well as the limits set by a global geopolitical structure. Countries are an example of just one geopolitical agent, comprised of others, and interacting with other countries, non-state organizations, and multiple-state organizations (such as NATO and the United Nations (UN)) within a geopolitical structure. The hierarchical nesting of agents and structures can be conceptualized as interlocked geographic scales. All structures and agents are dynamic, their form and purpose contested. Such contestation requires us to think about different expressions of power, such as military capability and patriarchal relations, and their connections, in addition to the manner in which they are made to appear "normal."

Organization of the book

The book begins with the introduction of a simplified model of global geopolitics. The book ends with a discussion of the complexity, or "messiness," of geopolitical conflicts given the multiplicity of structures and the multiple identities and roles of agents. The text assumes no familiarity with geopolitical terms and no prior knowledge of conflicts, past or present. As you progress through the book, try to make your own understanding of geopolitics more sophisticated by exploring how the different structures and agents introduced in successive chapters interact with one another. Also, be engaged with quality newspaper and other media reports of current events. Use the text and the current events to: (1) identify the separate structures and agents and then, (2) see how they are related to each other. In other words, allow yourself to explore the complexity of geopolitics as you work through the book and become familiar with a growing number of structures and agents.

Within the overarching idea of structure and agency the book will be organized in the following way. Chapter 2 concentrates upon the global scale, and provides a way of thinking of a dynamic global geopolitical context, the structure within which countries must operate. Chapter 3 focuses attention upon the scale of countries, especially the choices and constraints they face as geopolitical agents. The foreign policy that negotiates these choices and constraints is called a geopolitical code. Chapter 4 remains with the topic of geopolitical codes, but shows the importance of how they are justified or represented. The representation of geopolitical codes is important for a country, in order for its actions or agency to be supported rather than contested.

Chapter 5 addresses geopolitical agents that construct and contest the state scale, as we formalize our understanding of countries by introducing them as states, and discussing the related concepts of nation, nationalism, and nation-states. The ideology of nationalism and the geopolitics of separatism are topics discussed in this chapter. Nationalism is a collective identity creating the assumption of community at the national scale and the correspondence of that identity with the spatial organization of society into nation-states. The ideological maintenance of states through nationalism is complemented through their territorial expression. Chapter 6 addresses the geopolitics of boundaries and boundary disputes as the means of defining the geographic expression of states.

From focusing upon the geopolitical agency and structural context of states Chapter 7 introduces another geographical expression of power, networks. The expressions "global terrorism" and "globalization" are common contemporary understandings that politics involves the movement or flow of things across boundaries and into the jurisdiction of states. These flows are both legal and illegal. The flows are facilitated by networks, whether a terrorist or criminal network, on the one hand, or the network of global finance that switches huge amounts of money from financial market to financial market across the globe. In Chapter 7 we will focus upon the topic of terrorism.

The final chapter summarizes the identification of geopolitical structures and agents, but complicates the picture by showing how contemporary conflicts are usually a combination of the structures and agents that have been treated separately in the preceding chapters. The book concludes by challenging you to continue to explore the role of geography on causing, facilitating, and concluding geopolitical conflicts: both those ongoing and those yet to come.

Having read this chapter you will be able to:

- understand the concepts of place and scale;
- understand the concepts of structure and agency;

- be able to think about places in the world as being unique and interconnected;
- be able to think of current events occurring within a hierarchy of scales;
- be able to think of current events as being performed by geopolitical agents;
- begin to consider how the actions of geopolitical agents happen within structures;
- consider the multiple forms of power that underlie geopolitics.

Further reading

Agnew, J. (2003) *Geopolitics*, London: Routledge.

A more in-depth and theoretically sophisticated discussion of geopolitical practice and the way it has changed.

Cresswell, T. (1996) *In Place/Out of Place*, Minneapolis, MN: University of Minnesota Press.

Develops and exemplifies the politics of place and identity, or the political geography of inclusion and exclusion.

O'Loughlin, J. (ed.) (1994) *Dictionary of Geopolitics*, Westport, CT: Greenwood Press.

An excellent resource for clarifying geopolitical terminology and also provides brief discussions of many geopolitical thinkers.

Ó Tuathail, G., Dalby, S., and Routledge, P. (1998) *The Geopolitics Reader*, London and New York: Routledge.

A collection of short essays proving easy access to many of the authors and documents introduced in this text.

Staeheli, L.A., Kofman, E., and Peake, L.J. (eds) (2004) *Mapping Women, Making Politics*, New York and London: Routledge.

An excellent collection of essays describing the feminist approach to the topics of geopolitics and political geography.

Taylor, P.J. and Flint, C. (2000) *Political Geography: World-economy, Nation-state, and Locality*, fourth edition, Harlow: Prentice-Hall.

An introduction to world-systems analysis (discussed in Chapter 2) as well as the broad content of contemporary political geography.

SETTING THE GLOBAL GEOPOLITICAL CONTEXT

In this chapter we will:

- introduce a geopolitical model to provide an understanding of the global geopolitical structure;
- discuss the different components of this model;
- interrogate the validity of the model;
- note how the model is both similar and different to "classic" geopolitical frameworks;
- emphasize how we can use the model to provide a structure or context to understand geopolitical agency.

Traditionally, geopolitics has claimed to be able to paint neutral and complete pictures of "how the world works": what drives historical changes, what causes countries to fight, what determines whether a country will become a great power or not. The classical geopoliticians of the nineteenth and early twentieth centuries invoked a "God's eye view of the world," providing simple histories or theories that, they claimed, not only explained what has happened in the past, but suggested particular policies to inform the actions of their own country in a global competition with others (Parker, 1985). In other words, geopoliticians made dubious claims of historical and theoretical "objectivity" to support their own biased view of how their own country should compete in the world.

Such a view of geopolitics is no longer in vogue. Any claim to be able to "see" a pattern of global politics is immediately challenged as being limited and biased—rightly so—because it is situated knowledge. Instead, attention is drawn to how geopolitical agents make strategic choices, and how these are complicated by competing goals and changing circumstances. In other words, increasing attention is given to agency over structure. However, decisions are not made within a social and political vacuum. As discussed in the previous chapter, agents are both enabled and constrained by structures. Countries make geopolitical choices, to go to war for example, while considering the wider geopolitical context. For example, China's increasing political and economic power has led to greater influence within East and Southeast Asia. The Chinese government, on the other hand, is being very careful not to provoke the dominant world power,

Box 2.1 Competition between China and the US

"Chinese Move to Eclipse US Appeal in South Asia" claims the *New York Times* headline (November 11, 2004, p. A1) for an article by Jane Perlez talking of university students in Thailand choosing to learn Mandarin over English as a result of perceptions that China will be the dominant influence in the region soon. The Chinese have funded and built many language and cultural centers

> as part of China's expanding presence across Southeast Asia and the Pacific, where Beijing is making a big push to market itself and its language, similar to the way the United States promoted its culture and values during the cold war.

the United States (see Box 2.1). For a historic perspective, at the beginning of the twentieth century Alfred Thayer Mahan's concern was to create a geopolitical plan for the United States to increase its global influence without provoking the then global power, Great Britain. On the other hand, British geopolitician Sir Halford Mackinder's concerns at the beginning of the twentieth century focused upon maintaining Great Britain's preeminence in the face of a growing German challenge.

Both of these examples suggest that geopolitical decisions are made with an eye toward the global geopolitical context, and especially the ability of a dominant power to set the agenda. In this chapter, we will introduce a contemporary model of geopolitics to define a global geopolitical structure. The model is George Modelski's cycle of world leadership and provides a structure within which the actions of states and other geopolitical actors may be interpreted. We will see that this structure is dynamic and use it to discuss how the global geopolitical context frames the actions of different countries. Though the chapter ends with a guide to allow for critique of the model, it may be useful to provide some cautionary notes here. Modelski's model of geopolitics is *not* capable of *predicting* events. It is a historical model that interprets a wealth of historic data in a simplified framework. In other words, it is a *descriptive* model. Also, Modelski's model is useful, but only within certain parameters. His view of geopolitics is limited to conflicts between the major powers; smaller countries and geopolitical actors that are not countries are not included in his model. However, the model is useful for introducing the idea of a geopolitical structure and offering a context for current geopolitical events. We will discuss the pros and cons of the model in greater depth at the end of the chapter.

Defining a global geopolitical structure: using and interrogating Modelski's model of world leadership

Mahan and Mackinder, as well as Ratzel and Kjellen in Germany, exemplified the state-centric perspective of geopolitics, and geographical determinism (as discussed in

Chapter 1). From their perspective geographic size and location, and the internal make-up of a country determined power. Subsequent, and purportedly more scientific, calculations of power have rested upon the economic, military, and demographic elements of a particular country. To understand state power and global geopolitical context, however, these ingredients must be related to the ability of a state to define the global geopolitical agenda. In other words, the Gramscian notion of power within a country that we introduced in the previous chapter has relevance for global geopolitics. Following Gramsci, we would expect the most powerful countries to wield (or at least attempt to wield) an ideological power over the other countries: the most powerful country would try to set a political agenda that the rest of the world would, more or less, follow. Two theories have been particularly influential in the discussion of this type of global agenda setting: Wallerstein's concept of hegemony (see Box 2.2) and, the one we will engage, Modelski's (1987) concept of world leadership.

Modelski's model of world leadership is a historically based theory, founded upon his interest in naval history. Power, for Modelski, is a function of global reach—the ability to influence events across the world. In history, such power has required control of the oceans. Hence, for Modelski, world power rests upon the ability of one country to concentrate ocean-going capacity under its own control. Ocean-going capacity is measured by the combined tonnage of a country's military and merchant navies. In this sense, Modelski echoes Mahan's insistence on the important role of sea-power.

Box 2.2 Wallerstein's world-system theory

The sociologist Imanuel Wallerstein profoundly challenged modern social science through his concept of the historical social system. His argument was that society should not be equated with a particular country, but rather at a larger scale of the social system. According to Wallerstein, since approximately 1450 the social system has been the capitalist world-economy. Within this theory, primary geopolitical powers are called hegemonies or hegemonic powers. Since the twentieth century, the United States has acted as hegemonic power. The basis for hegemony is economic strength that translates into a dominant influence in global trade and finance. Maintenance of the capitalist world-economy in a form that benefits the hegemonic power requires, at times, military force. Hegemony is seen as an economic process for selfish goals, and not the global political benevolence of Modelski's world leadership. Similar to Modelski's model, the hegemonic power emerges from a period of global conflict, but Wallerstein is adamant that the United States is currently experiencing a relative decline in its global dominance. One other important difference is that in Modelski's model there is always a world leader, though its strength is cyclical. For Wallerstein, periods of hegemony are rare. So, if the US's hegemony does decline, according to Modelski a new leader should emerge after a period of war. Wallerstein's model suggests that other political scenarios, without one dominant state, may emerge.

However, most significantly, for our understanding of the contemporary world, world leadership is not defined solely by this material measure of power. Indeed, it is important to reflect upon the name Modelski gives to dominant and powerful countries—they are identified as world leaders, not hegemonic or superpowers. Remember, a crucial component of geopolitics is representation. Modelski portrays the world's most powerful country as a "leader," implying willing followers, rather than a hegemonic or superpower with its allusions to dominance and force.

Obviously, Modelski's definition of power is of the ilk that is strongly criticized by feminists (see Chapter 1). Power, in the model, is about strength and dominance, it is about the ability to exercise military force across the globe. This is another way in which Modelski follows the "classic" geopoliticians. This notion of power leads to an uncritical belief that the militarization of foreign policy is inevitable and beneficial. It also ignores gender relations within states and global economic inequities. In other words, Modelski's notion of power is uni-dimensional. We may still agree that a feminist critique of Modelski's power index is valid and yet still find value in the model. In fact, in the subsequent chapters we will see that geopolitics is represented in certain gender specific ways for the power relations Modelski identifies to be sustained. In other words, by bringing a feminist critique to bear upon Modelski we can get more out of the model than was originally intended by its author.

Activity

Is it more accurate to think of the United States from the early twentieth century to today as a hegemonic power or as world leader?

To answer the question, think about the relative weight given to economics versus politics and self-interest and political duty in the different models. Perhaps you may be able to find examples of both.

Is the global "responsibility" of the US as either world leader or hegemonic power solely a matter of rhetoric, or can you also point to particular actions?

A world leader is a country that is able to offer the world an "innovation" to provide geopolitical order and security. By innovation Modelski means a bundle of institutions, ideas, and practices that establish the geopolitical agenda for the world. The power of the world leader rests in its ability to define a "big idea" for how countries should exist and interact with each other; an idea that it is able to put into practice through its material power or naval capabilities. The power of the world leader rests in its agenda setting capacity and its ability to enforce it.

Modelski's model of world leadership is dynamic. The strength of the world leader rises and falls. Over the course of centuries, the mantle of world leadership has passed from one country to another in a sequence of cycles of world leadership (see Table 2.1). Each cycle of world leadership lasts approximately 100 years and is made up of four roughly equal phases of about 25 years (Figure 2.1).

Table 2.1 Cycles of world leadership

World leader	Century	Global war	Challenger	Coalition partners
Portugal	1500s	1494–1516	Spain	Netherlands
Netherlands	1600s	1580–1609	France	England
Great Britain	1700s	1688–1713	France	Russia
Great Britain	1800s	1792–1815	Germany	US plus allies
United States	1900s	1914–45	Soviet Union/ al-Qaeda	NATO/Coalition of willing

Source: George Modelski (1987) *Long Cycles of World Politics*, Seattle: University of Washington Press.

■ *Phase of global war*: The ability, or perceived right, to act as world leader is decided through a period of global war. The declining world leader is challenged by countries believing they should inherit the mantle. Coalitions are constructed and over the 25-year period, that may include a number of different wars and conflicts, one country emerges as having both the material capacity and ideological message to impose global order.

■ *Phase of world power*: Once victory has been achieved the geopolitical project of the new world leader is enacted. New institutions are established to apply and enforce the new agenda. On the whole, the new agenda is welcomed and followed.

■ *Phase of delegitimation*: At the outset of the establishment of a new period of world leadership, the imposed "order" is, overall, welcomed. But over time dissent grows. The benevolence of the world leader can be questioned; its actions seen increasingly as self-serving. Alternative agendas are given greater weight. The challenge to the world leader has begun, but the world leader is still relatively strong.

■ *Phase of deconcentration*: The challenges beginning in the previous phase become stronger. The world leader expends its material and ideological capacity in reacting to these challenges, making it weaker and more vulnerable to more attacks, in a spiral of challenge and reaction that leads to the phase of global war. Challenges are more frequently, but not exclusively, violent and organized campaigns. The world leader is called upon to react militarily, exhausting its material base of power and highlighting contradictions between its actions and its rhetoric. In combination, its legitimacy is increasingly questioned, and challenge intensifies.

As a concrete example, the war in Iraq has been represented by the United States as a mission in the name of "peace" and "humanity," as we will discuss later. As part of that military action, atrocities came to light that added fuel to the fire of those opposed to the US presence in Iraq. A prominent event was the abuse in Abu Ghraib prison that provided a sharp contrast between the representation of the occupation of Iraq as a "civilizing mission" of world leadership and actual events that challenge the leader's authority (see Box 2.4).

Using the ideal conceptual framework we have discussed, Modelski paints a particular picture of history—one defined by the cycles of world leadership. The role of

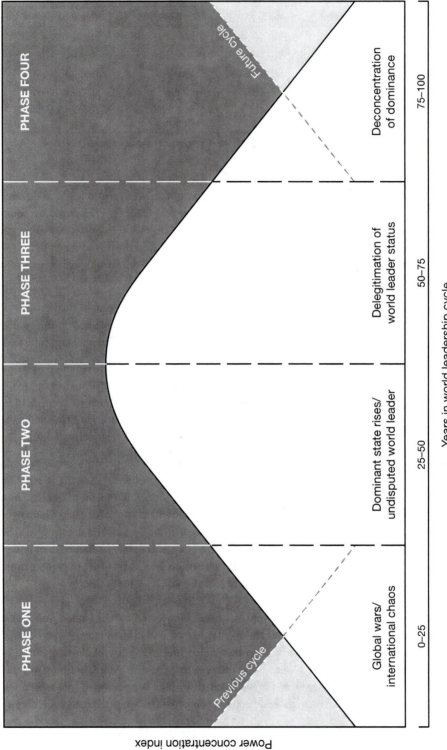

Figure 2.1 Modelski's world leadership cycle.

representation in his model is most important. In a cold use of language, global wars are defined as a "systemic decision"—they are instrumental in deciding who will be the next world leader (see Box 2.3). For Modelski, a leader is seen as acting benevolently—carrying the burden of maintaining global security for the benefit of all rather than acting for narrow national self-interest. The order defined by the "innovation" is portrayed as neutral; it is seen as being obviously good for all, rather than benefiting some countries or groups over others. Perhaps, most significant is the pattern of history Modelski identifies from the application of his model. Great Britain was able to have two consecutive cycles of world leadership. The geopolitics of the model is clear, if the Brits had two shots then there is nothing stopping the United States doing the same thing; the twenty-first century can be an American century too!

Box 2.3 World Wars I and II in historical context

Both Modelski and Wallerstein view the two world wars as twin episodes in one conflict, the one that decided who would succeed Great Britain as world leader/hegemonic power. Modelski is also guilty of representing these two (or is it one?) conflicts in cold language. Together, they are identified as a "systemic decision" of world leadership succession—a very instrumental way to view the deaths of millions of soldiers and citizens across the globe.

Within the phase of global war, the emerging world leader has a "good war," in the sense that it avoids much of the physical destruction of its homeland suffered by other fighting countries. Hence, its relative economic power increases dramatically. In the case of World War II, as the factories of Germany, Japan, and Great Britain were being flattened by aerial bombing, those in the United States were expanding their capacity. The emerging leader also enters the conflict relatively late—using its relative power to dictate the terms of peace to its likening. For further reading see Peter Taylor's use of Wallerstein's framework to analyze how Great Britain faced opportunities and constraints in creating its post-World War II foreign policy in his book *Britain and the Cold War* (1990).

The United States and Modelski's model

In the next chapter we will spend more time talking about how the geopolitics of the United States changed over the time periods of the cycle. For the moment, we can offer a sketch of American history to help you relate the abstract model to the "real world." The period of global war in this particular cycle ran from about 1914–45, the beginning of World War I to the end of World War II. The US played a minor role in the former conflict while it came in late and decisively in the latter. At the end of World War II, the US was able to set a global agenda around the twin themes of national self-determination and development that established its position as world leader. Institutions such as the International Monetary Fund (IMF), UN, and NATO were established to enforce and legitimate the new world leader's agenda. However, dissent toward the US's leadership emerged, and much quicker than in previous cycles. The Soviet Union provided an immediate ideological and military challenge. The Vietnam War exposed the world leader to allegations that it supported continued European-style control of the poor ex-colonies in the world, and illustrated the limitations of its military capabilities. The Korean War and the Vietnam War are evidence that the US suffered from violent coordinated military challenge much earlier than Modelski's model would suggest. As the twentieth century drew to a close, a different form of challenge emerged at about the same time Modelski would say the US was entering the phase of deconcentration. The anti-US terrorism of al-Qaeda had sporadic successes in Africa and the Middle East prior to the devastation of 9/11 and the heralded "War on Terrorism."

Broadly, the twentieth century history of the United States fits the pattern expected from Modelski's model. Though it is interesting to note that challenges to the United States' leadership came much earlier than expected, and it is a matter of both interpretation and geopolitical guesswork whether the "War on Terrorism" is a period of deconcentration preceding a new phase of global war, or the global war itself.

How can we interpret the Cold War within Modelski's model? On the one hand, the Cold War shows that the US was challenged strongly much earlier than Modelski's model would expect. The ideology of Communism, under the guise of Marxism-Leninism, offered an alternative to the liberal-capitalist model proposed by the world leader. The world leader was unable to extend its influence globally, being excluded from the Soviet bloc and facing competition from socialist movements in Africa, Asia, and the Americas.

A key event in the era of United States' world leadership was the demise of the Soviet Union and the collapse of the Iron Curtain. In a series of events through 1989 and 1991 that took commentators and policy analysts by complete surprise, the countries of Central and East Europe that had been under the control of the Soviet Union since the end of World War II were allowed to renounce the Communist system. Spontaneously, in 1989, the physical barriers of the "Iron Curtain," most notably the Berlin Wall, were torn down by the bare hands of jubilant people who were eager to make contact with the West. In 1991, the Soviet Union became Russia and spoke of creating a democratic political system with a market economy in place of a Communist one-party state. How should what commentators in the US interpret as the "victory over Communism" be interpreted? One argument is that the US's first cycle of world leadership was truncated and successful. The Cold War represents a victory in a Modelski style global war that has ushered in a second consecutive cycle of world leadership for the US, under the guise of President George H.W. Bush's "new world order." However, both the lack of overt conflict with the Soviet Union and the current challenges being faced by the US undermine an interpretation that we are within the US's second cycle of leadership.

Alternative views of the Cold War may help us interpret it within Modelski's perspective. For analysts such as E.P. Thompson (1985) and György Konrád (1984), the Cold War was a mutually beneficial geopolitical drama that served the Soviet Union and the US, rather than a potential global nuclear holocaust. The Cold War provided the grounds for both major protagonists to control their allies in Western and Eastern Europe respectively. It provided the reason for the military occupation of Europe by both the Americans and the Soviets. In addition, the Cold War included a consensus that the poorer parts of the world were to be dominated by the big powers. Though both sides claimed the mantle of anti-imperialism, the Cold War provided the excuse for political and military control of the newly independent countries.

The most likely interpretation is that the Cold War signified a limited but significant challenge to the US's world leadership. In other words, the period of world leadership was muted and the period of delegitimation amplified. The argument that the Cold War was of mutual benefit to the Soviet Union and the US is supported by an interpretation that the beginning of the period of deconcentration (and not a period of stability) was marked by the collapse of the Soviet Union. All of a sudden, the certainties that the world leader had known were gone, and a violent challenge, that was hard to pinpoint and counter, emerged.

The European Union and Modelski's model

How do we interpret the European Union (EU) within Modelski's model? First, the genesis of the EU was part of US plans to rebuild Western Europe after World War II. Though there have been political disagreements across the Atlantic since 1945, in general the US has supported the integration of Western Europe, because it helped counter the challenge of the Soviet Union and also reinforced economic and political ties with the world leader. The countries of Western Europe have, generally, followed the will of the US. One historic dispute was the British and French attempt to seize control of the Suez Canal in 1956. However, this episode met with strong US disapproval and Britain and France quickly complied with the world leader's wishes by retreating.

The EU is the product of a trend toward intensified integration of European countries, coupled with an expansion of the number of countries included. Now the EU contains the countries of Central and Eastern Europe that were once under the control of the Soviet Union. The intensification and expansion of the EU has resulted in discussions of its assumption of a global geopolitical role. In some cases this role has been visible, and in others it has been conspicuous by its absence. For example, the EU countries have been influential in international negotiations over global warming emissions. Alternatively, in the 1990s European countries stated that they would take the lead in resolving the war waging in the former Yugoslavia, but after embarrassing failures it was ultimately the US who intervened militarily and diplomatically.

On the one hand, the EU may be viewed as a form of delegitimation, in Modelski's terms. Its growing strength and confidence has allowed some countries, notably France, to be critical of US policy. Significantly, the EU has established a military force, the EuroCorp. This may also be viewed as delegitimation: it is a statement that NATO (the military expression of US influence over Europe) is no longer taken for granted and that purely European alternatives may one day replace the world leader's institution. In 1992, the EU described EuroCorps as "a European multi-national army corps that does not belong to the integrated military structure of the North Atlantic Alliance (NATO)."

On the other hand, the current EuroCorps website contains the sub-heading "A Force for the EU and NATO." In 1993, EU documents clearly identified EuroCorps role within both the geopolitical structures of the EU *and* NATO. In addition, when push comes to shove, the European countries have supported the global military role of the world leader; most notably regarding the US decision to invade Iraq. The *potential* of the EuroCorp to allow the EU to project military power independent of, and even against the wishes of the world leader is evidence of delegitimation. However, the subordination of EuroCorp within NATO, and the practical constraints on its ability to act independently of the US is evidence of the continued power of the world leader.

In summary, the current signals from the EU are mixed: there have been verbal protests against US actions. The political decisions of the EU and institutional developments such as EuroCorps may also be interpreted as discontent with the world leader's agenda. Additionally, there have been trade disputes between the EU and US. Significantly, however, the trade disputes and construction of EuroCorps do not undermine the general agreement over free-trade policies between the EU and US, nor the inability of the EU to define and execute militarily operations free from the world leader's agenda. The EU is still the world leader's key ally; though some would say an increasingly reluctant one.

The documents discussed above can be found at the EuroCorps website www.euro-corps.org/site/index.php?language=en&content=home, revised and accessed September 12, 2005.

The geopolitics of the rise and fall of world leaders: the context of contemporary geopolitics?

Modelski's model helps us to interpret the major contemporary global geopolitical issue: the attempt by the United States to maintain its preeminent power status in the face of challenges to its leadership. To do this we can consider the dynamics of two separate but related concerns. First, is there a country willing and able to act, or as Modelski may well say "serve," as world leader? In other words, is there an availability of order, the possibility of one country, the world leader, to offer and enforce a geopolitical innovation? Second, does the rest of the world, or at least a significant majority, want that order? In other words, is there a preference for the world leader's imposed order, or would countries rather face the "chaos" or "insecurity" of competing agendas? Note the role of representation here again, as "insecurity" and "security" are often based upon the degree of acceptance of the world leader's agenda.

For each of the four phases of a cycle, we can compare the balance of preference and availability of order (see Figure 2.2). In a period of global war, no one country is strong enough, relative to others, to establish a global geopolitical order. After the emergence of a world leader, there is a desire for order and the world leader's agenda is followed, more or less. By the next phase, delegitimation, the order being provided by the world leader is beginning to be questioned. However, the world leader still retains its relative power advantage, and hence challenge to the world leader rests, on the whole, in the realm of diplomatic and verbal protest, though some sporadic military resistance may be witnessed. During the deconcentration phase of the cycle, not only has dissent toward the world leader's order heightened, but also the world leader's ability to enforce

Modelski phase	Preference for world order	Availability of world order
Global war	High	Low
World power	High	High
Delegitimation	Low	High
Deconcentration	Low	Low

Figure 2.2 Preference and availability of world leadership.

its agenda has declined too. In this phase, there is an increased challenge to the world leader, not only in terms of diplomatic and political agendas, but also in the form of organized military challenges.

Imperial overstretch

Global opinion is only one factor in explaining the process of the decline of world leadership. Emphasis has also been placed upon the relationship between the demands placed upon the cost of the world leader's military and its economic strength, or the ability to pay. During the world leadership phase of the cycle, where the new global agenda is mostly accepted, enforcement can be attained by a global naval capacity—the strategy of gunboat diplomacy whereby the mere presence of the world leader's navy is enough to keep potentially dissenting countries in line. Such a strategy is relatively cheap as the very costly undertaking of protracted military conflict is largely avoided. However, as the cycle progresses, and challenges to the world leader's authority increase in frequency and intensity, then the world leader is drawn increasingly into conflicts on land (Figure 2.3). Associated rising costs further drain the world leader's power and invite more challenges. In addition, the ghastliness of warfare provokes specific incidents that are used by opponents to challenge the moral authority of the world leader (see Box 2.4). In other words, the resort to increased land conflict is costly in both economic and ideological terms.

Interpreting popular representations of US geopolitics within Modelski's model

The process of an increased need to fight land battles as the efficacy of naval presence decreases is known as imperial overstretch (Kennedy, 1988). This idea was very much in vogue in the late 1980s. However, the triumphalism of the "victory" of the Cold War soon replaced the doubts regarding the "relative decline" of the United States. In the language of President Reagan's politics, US military expenditure had bankrupted the Soviet Union and *not* the world leader. After the Cold War, and the geopolitical certainties that it had provided, global geopolitics had to be interpreted in new ways. In this section we will briefly introduce three writers who attempted to explain, and perhaps, influence, US foreign policy through the 1990s.

The first is Francis Fukuyama (1992) who, while employed by the State Department, argued that "history" had ended: meaning that all ideological conflict had been resolved because the superiority of liberalism was generally accepted. In the framework of Modelski's model, no delegitimation and deconcentration were going to take place. Fukuyama's notion that the US had triumphed to such an extent that no challenge was likely was short-lived. The doubters soon had the audience's ear. Robert Kaplan's contribution to the *Atlantic Monthly*, "The Coming Anarchy" (1994) and subsequent pieces were notifications that the world leader's agenda had failed in many parts of the world. Promises of development and national self-determination, the essence of the US's world leadership innovations, had not been fulfilled. People were playing by another set of political rules. Samuel Huntington's, a political scientist at Harvard University, provocative and empirically inaccurate, *Clash of Civilizations* (1993) thesis was directly opposed to Fukuyama's message, by mapping out the forthcoming phase of global war. Far from the US's period of leadership never-ending, Huntington was providing a call to arms by identifying potential challengers to world leadership.

Other authors were challenging the "benevolence" of world leadership. Michael Hardt and Antonio Negri's (2000) book *Empire* was one of numerous books and articles, primarily by critical academics, that identified the United States post-9/11 as an imperial power, controlling territory through its economic and military might for its own material self-interest. The beginning of the twenty-first century was designated as the new American and imperial century. Perhaps, these authors are correct.

Great Britain s military expenditures

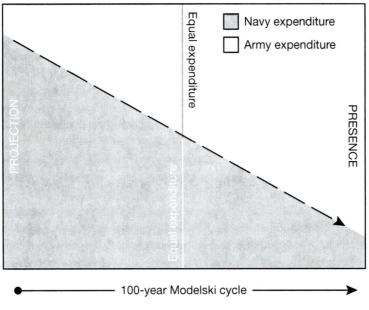

Figure 2.3 Imperial overstretch.

But Modelski's model cautions us to consider another possibility instead. Rather than the US's military might and activity being a sign of its imperial strength, the notion of imperial overstretch suggests that such developments are a sign of growing weakness. In this interpretation, challenge to the agenda of the world leader has become so intense that it must fight, and holes are being discovered in its military capacity and ideological shield.

Box 2.4 Abu Ghraib and the consequences for world leadership

In the spring of 2004 a series of photographs hit the global media that shattered the US's attempt to portray its occupation of Iraq as a humanitarian mission of the world leader intent on promoting human rights. Soldiers had taken pictures of practices in Abu Ghraib prison in Iraq of inmates being subjected to demeaning and painful acts, tantamount to torture, designed to break their resolve prior to interrogation (Figure 2.4). The images included those of a terrified inmate warding off a prison guard's attack dog, an inmate forced to kneel as if performing oral sex on another, laughing soldiers standing by a pyramid of naked detainees, inmates being led around on leashes by prison guards, and a hooded man posing as if on a crucifix with electric wires attached to his hands and penis.

In his investigation of the US abuses, Major General Antonio Taguba found practices that included:

> Breaking chemical lights and pouring the phosphoric liquid on detainees; pouring cold water on naked detainees; beating detainees with a broom handle and a chair; threatening male detainees with rape . . . sodomizing a detainee with a chemical light and perhaps a broom stick, and using military working dogs to frighten and intimidate detainees with threats of attacks, and in one instance actually biting a detainee.

Ideologically, Abu Ghraib was a disaster for the US's portrayal of itself as world leader, and will have lasting impacts. The photographs alienated politicians in foreign countries who supported the US's mission of creating democracies through military presence, or put them into a position where they could no longer support US actions because of negative public opinion. Obviously, the images inflamed those already opposed to the US's role in Islamic countries, and were used to justify their existing rhetoric of Americans as "infidels."

The abuse in Abu Ghraib, along with allegations of torture at Camp X-Ray, Guantanamo Bay, was not inevitable or predetermined. However, they were a product of the world leader's self-imposed policing mission and the dissemination of the images of torture undermined the ideological authority underlying its position. In other words, the actions of the world leader contradicted its rhetoric.

Figure 2.4 Camp Delta.

Evidence of imperial overstretch?

Representation is only one aspect of geopolitics. We can also look at what the US, as world leader, is actually doing; the policies it is forming, and interpret these through the lens of imperial overstretch. A continuous debate concerning the war in Iraq was the number of US troops needed for the initial invasion and the subsequent "post-invasion" goal of forming a democratic Iraq. Secretary of Defense Donald Rumsfeld has been continually criticized for not deploying enough troops. In the US military organization, any large military operation, such as the invasion of Iraq, relies upon the deployment of the National Guard, a military reserve of citizens. The deployment of these soldiers (men and women) has an impact upon families, communities, and businesses as mothers, fathers, neighbors, employees, etc. are sent overseas. The impact of fighting has long-term effects too, as recruitment into the military and the reserve may decline (see Box 2.5).

The dynamics of US troop deployment may also shed light upon the dynamics of world leadership (see Table 2.2). "The presence of American forces overseas is one of the most profound symbols of US commitment to allies and friends" (*Quadrennial Defense Review Report*, 2001, p. 11). Indeed. The demise of the Soviet Union and the identification of new challenges to US authority have initiated a rethink on where in the world US forces are to be located. There are plans to remove 50,000 US troops from Europe and 12,500 from South Korea (Garamone, 2004). The Pentagon has identified "weak states" in Asia, Africa, and the Western Hemisphere (which altogether is a huge chunk of the world) as the source of future challenges requiring the continued establishment of bases in Central and Southern Europe and Central Asia, while maintaining bases in Japan and Western Europe (*Quadrennial Defense Review Report*, 2001, p. 11). Changes are afoot, including talks of Stryker Brigade Combat Teams, which are more maneuverable than current brigades (Peltz *et al.*, 2003). These reorganizations suggest that in a time of challenge to its power, the US military is preparing for a fluid and less precise geography of threat than in the Cold War. Also, one of the (many) new buzzwords is "capability threat," whereby the military does not prepare for a known military threat (such as the Soviet Red Army) but an enemy whose identity and means of fighting cannot be predicted (*Quadrennial Defense Review Report*, 2001, p. 13). If these developments were to be interpreted through Modelski's model, it would appear that the US military was once organized to police an established "order," but in the face of challenge the nature and location of threat has become less predictable.

Table 2.2 US global troop deployments

United States	Europe	East Asia/ Pacific	North Africa, South and Southwest Asia	Sub-Saharan Africa	Western Hemi-sphere (not US)	Operation Iraqi Freedom	Other global forces
1,166,000 (normally)	114,300	96,900	5,900	900	1,700	167,300	37,300

Source: Kane (2004).

Box 2.5 Stress on the military power of the world leader —food for thought in the US military

The Pentagon's reliance on volunteers from the Army Reserve for duty in Iraq and Afghanistan risks creating a "broken force," the reserve force's commander warned his superiors in a December [2004] memo, and he urged a wider call-up of reservists to active duty.

In his memo, Lt. Gen. James Helmly stated that the Army Reserve is no longer able to meet its commitments in Iraq and Afghanistan, nor can it "reset and regenerate" units for future missions.

Reserve commanders spend too much time trying to accommodate troops who don't want to serve, leaving the force unable to meet its mission requirements, Helmly concluded—the result of policies that were designed for peacetime, "as opposed to a mobilized force in wartime."

"While some have expressed surprise and indignation at being mobilized for this war, most have not," Helmly wrote in a December 20 [2004] memorandum to Gen. Peter Schoomaker, the Army's chief of staff. "They have understood it to be inherent in their volunteer contract for service."

In addition, offering a $1,000 monthly bonus to volunteers for active duty risks creating a "mercenary" culture in its ranks, he wrote.

About a fifth of the Army's 200,000 reservists are currently on active duty. About 30,000 of those are in Iraq, where the service relies heavily on reservists for support, military police, and civil affairs specialists.

(CNN, 2005)

A related development was the US military's creation of Operation Blue to Green in the summer of 2004. The new program sought

> to recruit airmen and sailors leaving their service due to force reductions into the Army, which is temporarily reducing its ranks. Plans call for the Air Force to reduce its numbers by 16,000 and the Navy by 7,900 by the end of 2005, officials from the two services confirmed.
>
> (American Forces Press Service webpage, 2004)

The first story indicates stress on a vital component of the US's military, the Army Reserve as it is required to support operations in Iraq and other commitments, such as the Balkans. An increase in the need for "troops on the ground" is expected in the latter stages of Modelski's model and is the key feature of "imperial overstretch."

The second story is indicative of a movement of military resources from those services that project a global presence to those required to fight actual conflicts "on the ground." Again, it is precisely this process that Modelski identified at the end of previous cycles of world leadership, as increased challenge to the world leader required combat (see Figure 2.3).

Let us remind ourselves that Modelski's model is not a crystal ball. We cannot utilize its simplification of history to predict the future. We can use it to provide perspective upon current events. Is the idea that the US may suffer from imperial overstretch passé? If the United States was to follow the same cyclical pattern as previous world leaders, we would expect it to be an increasing problem.

Legacy, change, and world leadership: feedback systems in Modelski's model

The final feature of the model we will discuss is its feedback system. Modelski identifies two related feedback systems. The first, the developmental loop, notes that though the world leaders come and go the legacy of their innovation remains. In other words, the ideas and institutions established by the world leader do not disappear entirely from the geopolitical scene as a particular country loses its status as world leader. For example, if the US was replaced as world leader it is likely that the idea of national self-determination that was an ingredient of its "innovation" will still retain some role in global geopolitics. Also, the institutions of the UN and the World Bank, as entities managing global economics and politics are likely to remain, if perhaps in a different form. As support for this claim, the "ideas" of free trade and freedom of movement in international waters established by world leaders hundreds of years ago remain essential political norms.

The second feedback system outlined by Modelski is the regulatory loop that examines the process of an emerging challenger and the establishment of a new world leader. The logic of Modelski's model does not allow us to make predictions. It is difficult to consider this model without asking who will be the next challenger, and who will be the next world leader. Specific answers are not provided. However, recourse to Modelski's model does raise some interesting historical patterns that help us interpret the current situation.

In Modelski's history, the next world leader has not been the challenger, but has been one of the countries in the coalition brought together by the world leader to fight the challenger. The case of the United States and Great Britain is a clear illustration of this process. Great Britain's role as world leader was challenged by Germany resulting in the two world wars. To challenge the might of the world leader Germany realized it needed to form a coalition; it could not do it alone. However, given the process of decline identified by Modelski, Great Britain could not fight off challenges to its power alone either. It too needed to establish a coalition of forces. Crucially, it required the industrial might of the United States to support its war effort. Germany and Great Britain, challenger and leader, exhausted their material capacity for power in the long phase of global war. Remote from the domestic destruction suffered by Great Britain, continental Europe, the Soviet Union, and Japan, the United States gained ideological influence in relation to the relative and absolute increase in its material power. Both previous leader and challengers were spent forces, but the US, the increasingly prominent member of the world leader's coalition, was able to assume the preeminent geopolitical position. If there is a lesson to be applied from Modelski's model, it is that educated guesses about the next world leader should select from the coalition, the leader's allies, and not its challengers.

The current geopolitical situation complicates the ability to learn from Modelski's model, as will become clearer when we discuss the geopolitics of terrorism in Chapter 7. Modelski's historic examples are from the period when geopolitical actors were identified as competing countries. Other geopolitical actors were ignored. What of now? If we focus solely on countries, then China, the European Union, Japan, and to a lesser extent Russia are wheeled out every now and then as "threats" to the US. But, these countries are not the cause of the US's current military mobilization. Has the geography of challenge changed? Is the network of al-Qaeda the challenger to the US's world leadership? If so, what does that mean for coalition building and the process of succession?

Pros and cons of Modelski's model

Modelski's model is helpful for putting particular events into a historical perspective. Current affairs are not singular unrelated events. Rather, they are moments in broader processes and trends. Greater understanding of the event, its significance and implications, is achieved if you evaluate it within an understanding of world politics such as is offered by Modelski. Moreover, events can also be thought of as "observations" or "data." They are the "test" of the model. In other words, do the events we see in the news counter or support the trends we expect from Modelski's cycle of world leadership? Of course, the model must be thought of broadly and as an abstract teaching tool. Nonetheless, too many deviations from the expected pattern of events should lead us to challenge the model.

The model itself is far from perfect either. But that should not force us to dismiss its value out of hand. Social scientists are well aware that the theoretical tools we work with are imperfect. One of the most important concerns toward Modelski's model, and similar ones such as Wallerstein's world-systems theory, is philosophical. First is the logical problem of historical determinism. Just because Modelski has identified cyclical patterns of world leadership in the past, does not allow us to predict that the demise of the US's world leadership role is inevitable or determined. Portugal's sixteenth-century history does not determine the US's twenty-first-century future. The reason for this lies in another philosophical concern, structural determinism. The US as world leader is a geopolitical agent; it has some degree of freedom to choose its own actions. A drift toward global war is not determined; it *partially* rests upon the actions of the US.

The key word here is "partially." Proponents of structural models tend to give emphasis to the constraints that structures place on agents; in this case the structural inability of the world leader to untangle itself from increased challenges to its authority. Researchers who are more focused upon agency place greater emphasis on, say, the foreign policy decisions made by successive US presidential administrations.

Another set of criticisms toward Modelski's model reside within his conception of what is geopolitics. First, his model follows the classic geopolitical tradition of being state-centric. The geopolitical agents (leaders, challengers, and coalition members) are all countries. Second, he focuses upon the rich and powerful countries; poorer countries of the "global south" are deemed irrelevant in his system of challenge, war, and leadership. The geography of Modelski's geopolitics is limited in two senses; it sees state territoriality as the only space of politics and it concentrates on just a part of the globe.

Power is central to any understanding of geopolitics. Hence, we should be especially critical of Modelski's measure of power. One obvious question is whether sea-power is no longer relevant in an age of cruise missiles and satellite communication. In defense of Modelski, his long-term historical perspective requires a consistent measure of power, one that is as useful for understanding the sixteenth century as it is for comprehending the twenty-first century. Sea-power seems to fit the bill. The essence of the model, and the definition of power, is global reach, the ability to influence the behavior of other countries across the globe. At times this requires military muscle, and as we have seen in recent US-led conflicts that still requires a naval presence. Moreover, the US military has redefined the meaning of "global reach" utilizing weapons and surveillance systems that facilitate observation of the whole globe at all times, and the ability to remotely kill people and destroy targets across the globe (see Box 2.6). Unmanned drones carrying missiles and cameras may be a long way from sea galleons, but each identify the world leader as the country with the dominant means of exerting its power across the globe in their respective historical periods.

Also interesting to note is the contradiction in the measure of power and the operation of the politics of world leadership. While the power index is based upon a material measure, sea tonnage, the process rests upon ideological power, the ability of the world leader to define and implement a political agenda that is perceived to be in the interest of all. Rather than focusing upon the number of aircraft carriers and nuclear submarines, the world leader's authority rests upon the resonance of its political and cultural institutions and practices. The resort to arms is an admission that the political agenda is not being followed.

In subsequent chapters we will see how feminist and Gramscian notions of power can be applied to address questions of the representations and gender relations that are not only promoted by the power politics Modelski analyses but are, in fact, necessary for world leaders to maintain their authority.

Another pertinent question that is often raised centers upon the "driving force" underlying the model's dynamics. One attempt has been to relate the rise and fall of world powers to global changes in technology and economics (Modelski and Thompson, 1995). This raises the question of how to understand the material capacity and need to possess the sea tonnage that is an integral part of his model. Economic power requires a large merchant fleet to facilitate trade. Economic power provides the public funds needed to build a military naval capacity. In addition, Modelski forces us to look at some other processes. Most intriguing is the phasing of the preference for order. It implies a generational process of forgetting the horrors of warfare experienced by many during "global war" and an increasing truculence with an "imposed order": the geopolitics of mass psychology rather than the imperatives of capitalism.

Finally, is Modelski a geopolitician or a social scientist? A social scientist should be gathering and interpreting data with an eye to avoiding the biases of their social position and nationality. Geopoliticians, on the other hand, are politicians with an eye toward

Box 2.6 Technology and the global reach of the US military

Research and development efforts within the US are aimed at enhancing the technological capacity for the military's global reach (Figure 2.5). The Defense Advanced Research Projects Agency (DARPA)/Air Force Falcon program is developing hypersonic flying technology "that will enable prompt global reach missions and demonstrate affordable and responsive space lift" (DARPA, 2005). The unmanned reusable hypersonic cruise vehicle

> would be capable of taking off from a conventional military runway, carrying a 12,000-pound payload, and reaching distances of 9,000 nautical miles in less than two hours. This hypersonic cruise vehicle will provide the country with a significant capability to conduct responsive missions with quick turn-around sortie rates while providing aircraft-like operability and mission-recall capability.
>
> (DARPA, 2005)

In everyday language, this military robot can fly very fast, reach across the globe, bomb a target at a moments notice, and do it again soon afterwards, or be redirected while in flight.

A related form of global reach is Geospatial Intelligence (GEOINT), a form of military power in which the discipline of geography is heavily implicated. GEOINT is the "natural marriage" (National Geospatial-Intelligence Agency, 2004, p. 13) of satellite and rocket images with Geographic Information Systems.

> By combining remote sensing, precise geopositioning, digital processing, and dissemination, GEOINT enables combatant commanders to success-fully employ advanced weapons on time and on target in all-weather day-night conditions around the world. Today's warfighting capabilities represent quantum improvements in precision and targeting technologies.
>
> (National Geospatial-Intelligence Agency, 2004, p. 15)

And the goal? Well, "by continuing to leverage innovative technology and pro-cesses with an increasingly agile workforce, NGA (National Geospatial-Intelligence Agency) and NSG (National System for Geospatial-Intelligence) members are uniquely postured to contribute to information dominance and, ultimately, achieve the promise of a more certain world" (National Geospatial-Intelligence Agency, 2004, p. 17). Contemporary global reach requires "infor-mation dominance," the goal of "knowing" the world that the world leader dominates. The purpose of GEOINT is more "efficient" military operations that will facilitate a "certain" world: not necessarily "just" or even "peaceful" but "certain," a synonym for the "order" the world leader says it can provide to justify its relative power.

Figure 2.5 Unmanned military drone.

advancing a particular foreign policy agenda that they believe will enhance the interests of their own country relative to others. For geopoliticians, data are collected and theories are written in order to provide a seemingly objective backdrop that makes their political agenda seem "obvious" and validated by "science." Within which camp Modelski falls is a matter of interpretation. He does have a message for the geopolitical future of the US. He is also a skilled historical social scientist who has meshed an impressive data collection with an intriguing theoretical model.

Summary and segue

This chapter has introduced a particular model of world politics in order to set the global geopolitical context. Though Modelski's model is far from perfect, it does allow us to situate the actions of countries within a global picture of political cooperation and conflict. Perhaps the most important usage is in the interpretation of the role of the US, and why it appears to be facing increased and intensified opposition. As we shall see in the next chapter, despite its apparent focus upon the rich and powerful countries, it also helps us contextualize the geopolitics of al-Qaeda, with its violent opposition to United States' world leadership and rhetorical claims to represent the victims of US policy.

But now that we have introduced a way of thinking of a global geopolitical structure, the next chapter will focus upon the agency of an important set of geopolitical actors, countries.

Having read this chapter you will be able to:

- define the key components of Modelski's model;
- understand the critiques of the model;
- use the model to interpret current events, especially US foreign policy and reactions to it;
- use the model to interpret representations of US foreign policy by politicians, academics, and commentators.

Further reading

Bacevich, A. (2002) *American Empire: The Realities and Consequences of US Diplomacy*, Cambridge, MA: Harvard University Press.

An in-depth and accessible discussion of US foreign policy decisions since the end of the Cold War. The book provides a wealth of information that may be interpreted within Modelski's model, or used to evaluate the model.

Modelski, G. (1987) *Long Cycles in World Politics*, Seattle, WA: University of Washington Press.

The research manuscript that details the model used in this chapter and the historic data used to make the case.

Taylor, P.J. (1990) *Britain and the Cold War: 1945 as Geopolitical Transition*, London: Pinter Publishers.

Uses Wallerstein's world-systems framework to provide an accessible discussion of how Great Britain, a geopolitical actor, made foreign policy choices within the geopolitical context at the end of World War II.

Wallerstein, I. (2003) *The Decline of American Power*, New York: The New Press.

The "world-systems" take on the trajectory of the United States.

GEOPOLITICAL CODES: AGENTS DEFINE THEIR GEOPOLITICAL OPTIONS

In this chapter we will:

- introduce the concept of geopolitical codes;
- define the component parts of geopolitical codes;
- outline how geopolitical codes operate at different geographic scales;
- interpret the changing geopolitical codes of the US within Modelski's model;
- show that geopolitical actors other than countries also construct geopolitical codes by using the example of al-Qaeda.

At the very outset of talk of war upon Iraq in 2003, there was little doubt that Great Britain would be the United States' most active and loyal ally. This was not a matter of force. Tony Blair's government certainly had the choice to play a minor role in the conflict, or even try to use diplomacy to challenge President Bush's plan. Yet somehow it was "understood" that Great Britain would give political and military aid to its established ally. Tony Blair's decision illustrates the features of the geopolitical actions of countries that we will discuss in this chapter: a country may choose to make particular foreign policy decisions, these choices are limited to varying degrees, and a partial influence on the choices made is the history of allegiances and conflicts.

The previous chapter provided a means to understand the global geopolitical context. The goal of this chapter is to focus upon countries as geopolitical agents: the manner in which they make decisions within the global context. We continue the themes of geographic scale and structure and agency to interpret how countries make foreign policy decisions within regional and global contexts.

Geopolitical codes

The manner in which a country orientates itself toward the world is called a geopolitical code. Each country in the world defines its geopolitical code, consisting of five main calculations:

(a) who are our current and potential allies

(b) who are our current and potential enemies

(c) how can we maintain our allies and nurture potential allies

(d) how can we counter our current enemies and emerging threats

(e) how do we justify the four calculations above to our public, and to the global community

(Taylor and Flint, 2000, p. 62)

For example, Great Britain has defined its primary allies within the transatlantic and trans-European institutions of NATO and the EU. Furthermore, it has tried to retain influence across the globe through the establishment of the Commonwealth, made up of ex-British colonies. The latter has had mixed success, for example the expulsion of Zimbabwe from the Commonwealth for its brutal campaign against white farmers in the face of strong criticism from Britain. The identification of enemies is also dynamic. Almost overnight, as the Soviet Union became Russia, it quickly changed from intractable enemy to an ally.

Attempts to maintain allies take a number of forms (Figure 3.1, for example). Economic ties are one chief plank. The EU evolved out of relatively modest beginnings to integrate the economies of France and Germany to cultivate a peaceful Europe after the brutality of the two world wars. Cultural exchange is also another vehicle for maintaining or nurturing peace. Educational scholarships such as the Rhodes, Fulbright, and

Figure 3.1 US troops in Kosovo.

Goethe fellowships encourage international understanding and long-term ties. Business organizations such as the Rotary Club are also aimed at establishing linkages. The choice of "good-will" visits for incoming presidents and prime ministers is indicative of which international relationships are deemed most worthy of attention (Henrikson, 2005). For example, it is a tradition that the incoming US president meets with his Mexican counterpart at an early date.

Military connections are also seen as a means to maintain international cooperation. NATO is perhaps the strongest case, in which it is determined that an attack upon one member is considered an attack upon all. Another means of connecting with allies is the sale of military equipment that is expected to tie the, normally, weaker buyer to the more powerful seller. However, there is no guarantee of subservience. Weapons supplied to Iraq during its war with Iran were subsequently seen as threats by the sellers, the United States and Great Britain. Less overt, are the relationships fostered by military training (see Box 3.1).

Box 3.1 Power and the Royal Military Academy Sandhurst

The Royal Military Academy Sandhurst, the British Army's officer training school, has graduated officers from across the globe since the nineteenth century, and 3,500 Overseas Office Cadets have graduated since 1947, their website claims. The goal of cementing geopolitical relationships with the military and political elite of other countries is made quite explicit in the Academy's literature.

> Many overseas cadets have gone on to have distinguished careers in their own country. Some have become Head of State or head of government. But whatever their career paths, they have that brotherhood [despite the same webpage promoting its training of women officers] and sense of comradeship in arms built on the common experience of Sandhurst. The friendships that are forged here are important in peace and, increasingly, in conflict. Multi-national operations today draw in more countries than ever before.
>
> There are in training today sixty-six Overseas Officer Cadets from twenty-eight different countries. During the past five years the first cadets have come from Georgia, Latvia, Lithuania, and the Ukraine. Who will be next?
>
> To be Sandhurst trained is to join a club with world-wide membership.
> ("Overseas Cadets" Royal Military Academy Sandhurst,
> www.atra.mod.uk/rmas/courses/overseas.htm—accessed 1/14/05)

The Academy's website at www.atra.mod.uk/atra/rmas/ makes interesting reading, and is useful in exploring different types or expressions of power. What form of power can you discern from the text and images? Think especially of the Gramscian and feminist definitions of power from Chapter 1. To answer this question, think about what norms and values the website promotes.

Means to counter enemies are also varied. A once dominant but now, seemingly, outdated ingredient of the United States', Soviet Union's, and Great Britain's geopolitical codes during the Cold War was appropriately named MAD, for mutually assured destruction. Nuclear capability was strong enough to annihilate enemies many times over. Of course, most of this weaponry remains. The belief was that as destruction was assured, no one would dare start a nuclear war and "peace" would reign. At the other end of the spectrum is diplomacy; negotiations between governments to, at the least, prevent hostilities and, at best, nurture more friendly relations.

Sanctions are a common non-military means to force enemies to comply with ones wishes. An international campaign of sanctions and boycotts put pressure upon the South African government to end its apartheid policies. More recently, Iraq was targeted in a failed attempt to force Saddam Hussein to allow full inspection of his arsenal. Sanctions are often criticized for making the population suffer through lack of food or medical supplies rather than the politicians who formulate the policies in question. Countries can also change their opinion on the efficacy of sanctions; the British government under Margaret Thatcher disparaged the use of sanctions against apartheid South Africa, the governments of John Major and Tony Blair were strong advocates of sanctions against Iraq.

The fifth element of a country's geopolitical code (see Figure 3.2 and Box 3.2, for examples) should not be underestimated. The definition of an enemy, especially when it entails a call to arms, is something that can destabilize a government and lead to its fall. For example, in March 2004 Spanish Prime Minister Jose Maria Aznar was defeated at the polls in the wake of his support for the United States invasion of Iraq. Tony Blair expended political capital making arguments that support for President George W. Bush's War on Terrorism, and especially the invasion of Iraq, was essential for British security. In addition, intensifying the EU in the name of European peace and prosperity has proved equally exhausting for British governments.

Representational geopolitics is the essence of the fifth element of a geopolitical code. If enemies are to be fought, the basis of the animosity must be clear, and the necessity of the horrors of warfare must be justified. Enemies are portrayed as "barbaric" or "evil," their politics "irrational" in the sense that they do not see the value of one's own political position, and their stance "intractable," meaning that war is the only recourse. As we will see in the next chapter, these representations are tailored for the immediate situation, but are based upon stories deposited in national myths that are easily accessible to the general public.

Scales of geopolitical codes

Every country has a geopolitical code. For many countries their main, if not sole, concern is with their immediate neighbors: are they friends or enemies, is increased trade or imminent invasion the issue? But some countries profess to develop a regional geopolitical code in which they have influence beyond their immediate neighbors. China's calculations toward expanding influence in Southeast Asia are a good example, as is Egypt's role within the Arab world (Taylor and Flint, 2000, pp. 91–102).

Figure 3.2 British World War II propaganda poster.

Finally, some countries purport to have global geopolitical codes. The world leaders we identified in the previous chapter are the primary agents in this role. A challenge to their authority anywhere on the globe requires a response, for their legitimacy is based upon their global reach (Flint and Falah, 2004). On the other hand, world leadership requires world "follow-ship." Much diplomatic energy is spent to make sure countries are "on-board" the world leader's agenda. Any attempt by another country to create a global geopolitical code is interpreted as a challenge to the world leader. The influence of the Soviet Union within Africa, the Caribbean, Middle East, and Asia during the Cold War is an historical example of how combat was part of two competing global geopolitical codes (Halliday, 1983).

Though we can distinguish the power and influence of a country through designating its geopolitical code as local, regional, or global, it is false to separate local geopolitical codes from the global geopolitical context. Though the range of geopolitical calculations may be local, the influence of the global geopolitical context remains. For example, Hungary's decision to join NATO involved calculations about ethnic Hungarians in neighboring countries, and a future threat from Russia, but was still framed within the global authority and agenda of the United States (Oas, 2005). Hungary saw the changes in the global geopolitical context, as the world leader exercised its authority in Europe with the collapse of the Soviet Union, as an opportunity to advance its own security. The same idea can be applied to the way the "stans," the republics of Central Asia, have utilized the War on Terrorism to obtain military aid from the US.

Box 3.2 "Mapping the Global Future": the definition of threat

In January 2005, the National Intelligence Council, a group of "analysts" who synthesize "expert" advice and report to the head of the CIA, released a report entitled "Mapping the Global Future: Report of the National Intelligence Council's 2020 Report." This report is not a geopolitical code in itself, as it provides scenarios based on "intelligence" rather than policy itself. However, it provides a basis of authority, the type of power that is viewed critically from Gramscian and feminist perspectives, which will likely underlie the revision of the US's existing code.

The report is notable for its emphasis upon economic change, especially the growth of China and India and the implications for trade and oil consumption. As a result, perhaps we may see a return to the construction of Asia as an economic threat to US power that was predominate in the 1990s. Furthermore, terrorism inspired by Islamic fundamentalism is identified as a continuing threat. The US remains the dominant superpower, according to the report, but its influence is partially diminished as European integration solidifies and the voice of other countries in international institutions increases. Instability in the Middle East, Sub-Saharan Africa, and Latin America is also raised as a concern.

Perhaps as interesting as the predictions the report contained was the manner in which they were presented. The report received a lot of attention in the US press, with syndicated reports appearing in local newspapers. Particularly striking was the use of fictional scenarios within the report to give it urgency and accessibility. Two such scenarios were a fictional text conversation between two arms dealers and the fictional diary of the UN Secretary-General in 2020, which refers to horrific terrorist attacks in Europe in 2010 inspiring the Europeans to enthusiastically back the US War on Terrorism.

Though these scenarios are within the realms of possibility, their sophisticated artistic representation blurs the boundaries between fiction, entertainment, and objective analysis. The scenarios may drive policy, whether a series of large terrorist attacks occurs in Europe or not. In this document, the means of representation and the identification of the threat are blurred into a fictional presentation of potential threats and enemies.

The report is accessible at www.cia.gov/nic/NIC_globaltrend2020_s2.html. Reading the report allows for the consideration of many of the concepts we have discussed so far. How can this document be interpreted within Modelski's (1987) model? In what way does this report exemplify a focus on particular power relations and agents that feminists would criticize? From a Gramscian perspective, what is "taken for granted" in this report and what are the implications? Is the representation of the report's findings successful in justifying its content?

Activity

Many countries post key foreign policy documents on the web.

- Choose a country of interest to you and see to what extent their code is aimed at the local, regional, and global scales and how they are connected.

- Also, identify how both force and diplomacy are combined in the statements.

A do-it-yourself case study: decoding the geopolitics of Central Asia

During the Cold War period and prior to its break-up the Caspian Sea basin and Central Asia were firmly under the geopolitical control of the Soviet Union. However, the past couple of decades have witnessed related changes that now make this region an arena of geopolitical contestation (O'Lear, 2004). Oil and gas reserves in the Caspian Sea basin are viewed as essential components of world trade, especially as economic growth in China and India drive-up global demand (Figure 3.3). Yet, the size and accessibility of the reserves are disputed. Moreover, there is much geopolitical dispute over pipeline routes to transport the products through the politically volatile Trans-Caucasus region consisting of Georgia, Armenia, and Azerbaijan; the battle being over whether a route favoring European or Russian control is built (a geopolitical concern brought to the general public's attention via the James Bond movie *The World is Not Enough* (Dodds, 2003)).

In addition, the recent establishment of the "stans" (Kazakhstan, Krgyzstan, Tajikistan, Turkmenistan, and Uzbekistan) has added further political tensions. Kazakhstan and Turkmenistan have their own oil and gas reserves, but these countries are also seen as pivotal in the "War on Terrorism" as they are portrayed as a "battleground" between Islamic fundamentalist groups and the establishment of "free institutions," meaning nominal democratic practices (though corruption and lack of political openness are rife) and free markets.

The final ingredient in the mix is the role of three powerful countries, each seeking to play a role in the region: China, Russia, and the United States. China and Russia claim "traditional" or "established" influence in the region. In other words, they resort to previous geopolitical codes to justify contemporary ones. For the United States, its increased presence in the region is couched in terms of the War on Terrorism, with access to oil and gas underlying the concern.

Political dynamics within countries after gaining independence, religious and political agendas, oil and gas, and the presence of three countries with a history of antagonism combine to make a complex situation.

Below are extracts from media reports on the geopolitical codes of China, Russia, and the US as they pertain to the republics of Central Asia. Read the extracts with an eye to the five elements of geopolitical codes.

Figure 3.3 The Caspian Sea.

Sources: US CIA; O'Lear (2004).

CASPIAN SEA OILFIELDS

1	Kashagan	13	27th May	24	Bakhar
2	Kurmangazy		Guneshli	25	Kurdashi
3	Khvalynskaya	14	Chirag	26	Khamandag
4	Tsentralnoye		Kaverochkin	27	Araz
5	Apsheron Island		Dostlug	28	Lerik
6	Apsheron Bank	15	Azeri	29	Garasu
7	Darvina Bank		26th Baku	30	Shirvan
8	Artem Island	16	Pricheleken	31	Bulla-Deniz
	Pirallkhi	17	Zhdanov	32	Alyaty-Deniz
9	Gyurghang-	18	Lam Bank	33	Bulla Island
	Deniz	19	Gubkin Bank		Duvanny-Deniz
10	Zhiloy Island	20	Livanov		Sangachaly-Deniz
11	Gryazevaya Sopka	21	Kapaz	34	7th March
	Palchyg Pilpilas		Promezhutoch	35	Lokbatan-More
	Neft Dashlary		October Revolution	36	Gum-Deniz
	Neftian ye Kamni	22	Yuzhnaya 1&2	37	Peschanyy More
12	Azi-Aslanova	23	Shah-Deniz		

Extract 1

August 16, 2004, President George W. Bush introduced a radical plan to re-deploy 60,000–70,000 US troops from European and Asian bases staffed for the Cold War to countries in Central Asia. A "senior defense official" was quoted as saying: "In the case of Uzbekistan, we have cooperation with them today on the war on terrorism. And we have believed that the war on terrorism will be with us for a period of time. And the kind of cooperation that develops further with Uzbekistan and others in Central Asia really depends on those countries to the extent they want to work with us."

There were both positive and negative reactions to this announcement. According to some it would inflame anti-American sentiment in the region, for others the potential redeployment was linked positively to the possibility of economic and political reform.

("US Redeployment Plan Sends Ripples through Central Asia,"
CDI Russia Weekly, August 24, 2004.
www.cdi.org/russia/320–16.cfm—accessed January 14, 2005)

Extract 2

Central Asia and the Caucasus are emerging as the new focal point of rivalry between Russia and the United States in the wake of the Iraqi crisis. At the heart of the new standoff are rich oil and gas resources in the Caspian Sea basin, which may hold 100 billion barrels of oil alone. Washington has already a firm foothold in the local hydrocarbon industry, with US and joint US-British companies controlling 27 per cent of the Caspian's oil reserves and 40 per cent of its gas reserves.

In the post-Iraq scenario, the US has moved to put a bigger foot in the South Caucasus and Central Asia. It has mounted a titanic effort to revive GUUAM, a moribund economic and security group of five former Soviet states, Georgia, Ukraine, Uzbekistan, Azerbaijan, and Moldova, set up as a counterweight to Russian influence.

In Central Asia, the Pentagon has recently assumed talks with Tajikistan on the lease of three military bases in addition to the two the US established in Uzbekistan and Kyrgyzstan in the wake of the 9/11 attacks. Last year [2002], Tajikistan received $109 million in economic aid from the US and accepted its offer to renovate a runway at Dushanbe airport.

("A new Big Game in Central Asia" by Vladimir Radyuhin
CDI Russia Weekly, from *The Hindu* posted July 18, 2003.
www.cdi.org/russia/268–12.cfm—accessed January 14, 2005)

Extract 3

Quoting a report by the Institute of Foreign Policy Analysis: "The United States should not allow itself to sort of get baited into a Cold War or a Great Game perspective on its relationship with China and Russia," Sweeney says. "Those states have legitimate interests in seeing Central Asia stabilized and in defeating

Islamic extremism. So as long as their actions don't conflict with our core objectives in the war on terrorism, we don't need to be overly suspicious or reactionary to Russian or Chinese moves in Central Asia.

"Last October [2003], Russia opened an air base at Kant in Kyrgyzstan to provide an air component for a rapid deployment force that will operate under the aegis of the Collective Security Treaty Organization (CSTO). The CSTO is a partnership among Russia, Armenia, Belarus, Kazakhstan, Kyrgyzstan, and Tajikistan.

"Last year, members of the Shanghai Cooperation Organization—which groups China, Russia, and four Central Asian republics—held joint military exercises in Kazakhstan and China.

"In 2002, China and Kyrgyzstan conducted a joint military exercise on the border areas of the two countries."

("Central Asia: Report Calls On US to Rethink its Regional Approach" by Antoine Blua RFE/RL *CDI Russia Weekly*, February 13, 2004. www.cdi.org/russia/13feb04–13.cfm—accessed January 14, 2005)

Extract 4

Visiting the Tajik capital Dushanbe last month [October 2004], Russian president Vladimir Putin surprised his audience by pledging substantial financial investment in the Central Asian republic.

"The Russian side—both its state structures and private companies—intends to invest some $2 billion in the Tajik economy within the next five years . . ." Putin said.

And Tajikistan was not the only Central Asian country to receive promises from Moscow in October. Uzbekistan and, to a lesser extent, Kyrgyzstan, have been promised investment by Russia.

("Central Asia: Russia Comes on Strong (Part 1)" RFE/RL *CDI Russia Weekly*, November 19, 2004—accessed 1/14/05. www.cdi.org/russia/331–17.cfm)

Activity

From the extracts, pick out the web of alliances that is being created in Central Asia and the potential tensions. As a starting point, one may want to look at the actions of Kyrgyzstan and the potential tensions between China, Russia, and the US.

- What means of generating allies are being used in this situation?
- In what way is the context enabling and constraining the actions of the countries involved?
- In what way is Modelski's (1987) model of world leadership useful in considering the structure or context?
- In what way is the model useful in explaining the geopolitical dynamics in the Caspian Sea basin?

Geopolitical codes and the dynamics of world leadership

The global geopolitical context is the aggregation of the geopolitical codes of all countries. Of course, some geopolitical codes are more influential than others; the United States' actions define the global political scene to a greater degree than those of, say, Belize. If Modelski's model is correct, during a phase of world leadership the geopolitical agenda of other countries will be most strongly influenced by the world leader. It follows that the model predicts phases of deconcentration and global war to be times when the ability of other countries to pose geopolitical questions increases. By placing the geopolitical codes of particular countries within the phases of Modelski's model we can have an understanding of the opportunities and constraints that the global geopolitical context defines.

The period of global war that ended in 1945 was waged between the declining world leader, Great Britain, and its chief challenger Germany. However, as we noted in the previous chapter, it was Britain's key ally, the United States, that claimed the mantle of world leader. In this period, we can distinguish three important geopolitical codes: Great Britain's attempt to arrest decline; Germany's attempt to defeat Britain; and the United States' realization of world leadership.

Geopolitical codes of declining world leadership and challenge

British geographer and geopolitician Sir Halford Mackinder, introduced in Chapter 1, was intent in maintaining his country's supremacy, he identified a German threat, and thought the maintenance of the British Empire was the appropriate means. In his mind, the British Empire would secure Britain's economic power, and its global presence. Mackinder is noted for his identification of conquest of Eurasian "Heartland" as a means to control the world. He feared that Germany was utilizing the "new" railway technology to mobilize resources in Eurasia and challenge Great Britain. Mackinder's response was "sea-power," or more accurately a system of global reach built upon empire that required a large navy.

Senior British politicians were also able to make the linkage between control of Europe and world domination. In practice, their geopolitical code, in an echo of Mackinder's concerns, required a united empire and battleships. In a 1911 conference on imperial defence, Sir Edward Grey (Foreign Secretary) stated:

> So long as maintenance of Sea Power and the maintenance and control of the sea communications is the underlying motive of our policy in Europe, it is obvious how that is a common interest between us at home and all the Dominions.
>
> (Quoted in James, 1994, p. 342)

His audience agreed, and there was cross-party support for a campaign to build four Dreadnought battleships a year. The naval arms race between Great Britain and Germany saw plans for Germany to construct 61 battleships between 1898 and 1928.

The material aims of maintaining the British Empire to secure world leadership were represented in a way that emphasized a civilizing mission. War was a personal and national "duty" couched in religious language—personal loss was for a higher goal than national self-interest: in 1911 a National Service League pamphlet proclaimed "war is not murder, as some fancy, war is sacrifice—which is the soul of Christianity," and, remarkably, "fighting and killing are not of the essence of it [war], but are accidents" (quoted in James, 1994, p. 334). In other words, when the geopolitical code of the declining world leader required preparation for war, it required a representational politics of sacrifice for the benefit of all humanity. Walk around an old British church, and read the gravestones and plaques on the wall—one can see the loss of life suffered by the population of the world leader memorialized, and the losses given meaning in a mixture of nationalist and religious sentiments. Of course, what of the greater suffering of those subjugated under British imperialism? Representations of such losses are less readily available to most of us, producing a slanted and partial understanding of the "costs" and experiences of British world leadership.

The geopolitical code of a challenger is, by definition, aggressive. Germany felt "under-sized"; it did not possess the territory that it deserved given the status of the German culture. The expansionist geopolitical code of the Kaiser Wilhelm II was given a theoretical basis by the geopolitical theories of Rudolf Kjellen and Friedrich Ratzel (also introduced in Chapter 1). They emphasized the organic nature of the state, the state was an organism that would "naturally" grow (or increase the extent of its boundaries). Which state would control which territory was also justified through a biological analogy—the geographic realm. In other words, the "superior" culture would best manage the land and hence have the geopolitical right to possess it. Ratzel's allusions to American seizure of "Indian" territories was translated into the, for him, obvious benefits of German culture controlling the Slavic lands of Eastern Europe, to increase the *lebensraum* (or living space) of the German people (Smith, 1986).

But territorial expansion within Europe was not the only element of Germany's geopolitical code. The acquisition of new colonies was also paramount, hence the need for a new fleet of battleships. Germany believed that it was "owed" colonial possessions to the same degree as Britain and France, and the new emergent industrialists in Germany promoted this policy of *Weltpolitik* in its attempt to secure markets and resources. However, it should be noted that geopolitical codes are contested. Germany's landed aristocracy favored the acquisition of new agricultural land through expanding Germany's European borders through policies of *lebensraum*, while the industrialists favored the *Weltpolitik* strategy of securing colonies (Abraham, 1986). The twin aspects of Germany's geopolitical code of world leadership challenge enabled Great Britain to secure a coalition of support both across its Dominion possessions as well as traditional European rivals France and Russia.

Geopolitical codes of the US as world leader

But what of the emerging, though not pre-determined, world leader across the Atlantic? Recovering from a bloody civil war, impressively urbanized and industrialized in some parts, and "undeveloped" in others, the notion of expansion became a key issue in American geopolitics. Despite much political debate, control of the Caribbean and the Pacific became the focus of the United State's geopolitical code. Rear-Admiral Alfred Thayer Mahan was the theoretical light behind the US's move to globalism. Especially, he noted that sea-power was the basis for world power, but was also careful to caution that any expansion of US influence would have to be done in a way that did not interfere with Great Britain's agenda and provoke war.

US national ideology was, and still is, based upon the rhetoric of anti-colonialism and national self-determination from British rule. Hence, especially at the beginning of the process of achieving world leadership, there was much domestic accusation that the country was embarking upon a policy of European-style imperialism unsuitable for the United States. But expansion did follow, and key geopolitical achievements were the defeat of Spain, control of Cuba, Hawaii, and the Philippines, and the construction of the Panama Canal. Related was reinforcement of the Monroe Doctrine that defined the US's hemispheric sphere of influence across Central and South America, but also delineated, at a time of world-leadership transition, that Great Britain and the United States each had distinct and exclusive realms of control across the globe (Smith, 2003).

Such geographic limitations were unsuitable ingredients for a geopolitical code once the United States had achieved world leadership status. In the face of the ideological and territorial challenge of the Soviet Union, the new world leader had to create an unabashedly global geopolitical code. Table 3.1 illustrates how in 1947 the United States was including countries across the globe in its geopolitical calculations. In addition, policy toward particular countries was a function of "national security" and the new world leader's "mission" to counter Communism.

NSC-68, written under the administration of President Harry S. Truman in 1950 is the key document outlining the geopolitical code of United States world leadership (the National Security Council was established by President Truman to serve as a forum to advise the president on foreign policy). It is useful in showing the geographic imperatives a world leader must address, as well as showing the similarities with the foreign policy of President George W. Bush, when the world leader was facing a challenge from organized terrorism.

NSC-68 outlines the goals of world leadership, but it had to do so in the face of the geopolitical challenge of the Soviet Union and its ideological alternative, Communism. The document is oft-quoted for its claim that "[t]he assault on free institutions is world-wide now, and in the context of the present polarization of power a defeat of free institutions anywhere is a defeat everywhere" (Section IV, A). The geopolitical implication of this statement is that all parts of the globe held equal strategic importance—the world leader had to assert its authority in all countries. The Soviet system was a value system "wholly irreconcilable with ours" (IV, A), and its influence was preventing the establishment of "order" in the international system. Remember that "order" is a key component of Modelski's model: in the words of NSC-68, the conflict with the Soviet Union "imposes on us, in our own interests, the responsibility of world leadership" (IV, B).

Table 3.1 Constructing a geopolitical code for world leadership

Identifying the global mission		Adding the national interest	Ranking the world
Threat from Communism		US security	Prioritizing countries
9	Great Britain	1	1
6	France	2	2
U	Germany	3	3
3	Italy	7	4
1	Greece	10	5
2	Turkey	9	6
7	Austria	6	7
U	Japan	13	8
10	Belgium	4	9
12	Netherlands	5	10
17	Latin America	11	11
U	Spain	12	12
5	Korea	15	13
U	China	14	14
13	Philippines	16	15
18	Canada	8	16
4	Iran	U	U
8	Hungary	U	U
11	Luxembourg	U	U
14	Portugal	U	U
15	Czechoslovakia	U	U
16	Poland	U	U

Source: The data is from a Joint Chiefs of Staff document reproduced in Etzold and Gaddis (1972, p. 79 and pp. 82–3), and the table is slightly modified from Taylor (1990, p. 16).

Note: U = unranked.

Activity

As an aside, how should we evaluate the overlap in language between NSC-68 and Modelski's model?

- Consider whether it shows that Modelski was creating a theory to justify US foreign policy, in the classical geopolitical sense, or whether it illustrates his scholarly perceptiveness in identifying the essentials of world politics.
- Though there are no clear "right" or "wrong" answers, you may use this exercise to evaluate the goals and content of Modelski's model.

The geopolitical role of the United States as world leader was made clear:

> Our overall policy at the present time may be described as one designed to foster a world environment in which the American system can survive and flourish. It therefore rejects the concept of isolation and affirms the necessity of our positive participation in the world community.
>
> (VI, A)

In other words, the "American system" was the basis for the United States' role as world leader, and it required a global geopolitical code. The enemy was identified as the Soviet Union. Allies were countries and people advocating "free institutions."

The means of the geopolitical code were twofold. First, NSC-68 claimed a "policy to develop a healthy international community" (VI, A); the establishment of global order in Modelski's terms. Second, the document outlined a "policy of 'containing' the Soviet system" (VI, A), or negating the ideological and geopolitical challenger. Containment was a policy

> which seeks by all means short of war to (1) block further expansion of Soviet power, (2) expose the falsities of Soviet pretensions, (3) induce a retraction of the Kremlin's control and influence, and (4) in general, so foster the seeds of destruction within the Soviet system that the Kremlin is brought at least to the point of modifying its behavior to conform to generally accepted international standards.
>
> (VI, A)

As world leader, the US would be the influential investigator, judge, and jury when it came to breaches of "international standards," but this policy manifested itself in realms of activity from nuclear deterrence, to the Vietnam War, and espionage. There is a contradiction within NSC-68. On the one hand, it calls for the global role and presence of the United States, while, on the other hand, its call for "containment" acknowledges the challenge of the Soviet Union. In other words, the rhetoric of leading the whole world was maintained within the practical constraints of a bi-polar world.

But what of representing this geopolitical code of world leadership to domestic and international audiences? For domestic consumption, NSC-68 was based upon the ideals and content of the US Constitution. Section II was entitled "Fundamental Purpose of the United States" in which the "three realities" of individual freedom, democracy, and determination to fight to defend the American way of life were established and deemed to be under the protection of "Divine Providence." It was these "realities" that formed the basis of US world leadership; they were to be diffused to the world to maintain order. Section III, "Fundamental Design of the Kremlin," ("Design" having an evil, even sexual, implication rather than the valiant "Purpose") argued that the United States was the Soviet Union's "principal enemy." Both the domestic security and global mission of the US were justified by rhetoric within NSC-68. Hollywood was implicated too, as a spate of movies based on biblical epics portrayed the Middle East in a manner that was accessible while subtly justifying US foreign policy in the region. We will discuss these movies in greater depth in the next chapter.

Within the process of world leadership, geopolitical codes must also be dynamic. To illustrate this point we skip to another phase of Modelski's model, the contemporary period when the world leader must reassert its authority in the face of a new challenge. We will investigate the geopolitical code of the world leader through a discussion of the National Security Strategy (NSS) of 2002, the so-called "Bush Doctrine." First we will analyze the geopolitical code of what, at the time of writing, is the clearest geopolitical challenge to the US: Osama bin Laden's *fatwa*, or decree.

A geopolitical code to challenge the world leader

In February 1998, the London based Arabic language newspaper *al-Quds al-Arabi* published a statement signed by Shaikh Usama Bin-Laden, and four other men prominent in radical Islamic politics. The statement opened with two quotes from the Koran before setting the geopolitical scene for its readers:

> The Arabian peninsula has never—since God made it flat, created its desert, and encircled it with seas—been stormed by ant forces like the Crusader armies spreading in it like locusts, eating its riches and wiping out its plantations. All this is happening at a time in which nations are attacking Muslims like people fighting over a plate of food. In light of the grave situation and the lack of support, we and you are obliged to discuss current events, and we should all agree on how to settle the matter.
>
> (Quoted in Ranstorp, 1998, p. 328)

The "current events" were portrayed as "three facts that are known to everyone" (Ranstorp, 1998, p. 328). The overarching theme was the "self defense of Muslims against aggressive forces," manifest in the US military presence in Saudi Arabia since the first Gulf War in 1991, and the cooperation of the Saudi regime with the United States (Ranstorp, 1998, p. 328). The *fatwa* was a call to arms, and such a "fact" was portrayed in vitriolic language:

> [F]or over seven years the United States has been occupying the lands of Islam in the holiest places, the Arabian peninsula, plundering its riches, dictating to its rulers, humiliating its people, terrorizing its neighbors, and turning its bases in the peninsula into a spearhead through which to fight the neighboring Muslim peoples.
>
> (Ranstorp, 1998, p. 328)

This first "fact" was supported by two others; "the Americans are once again trying to repeat the horrific massacres" (Ranstorp, 1998, p. 329), in other words conflict with Iraq and other Muslim countries was about to intensify, and "the American's aims behind these wars are religious and economic, [but] the aim is also to serve the Jew's petty state and divert attention from its occupation of Jerusalem and murder of Muslim's there" (Ranstorp, 1998, p. 329). Part of bin Laden's success has rested on the second "fact," continuing through the 2003 war on Iraq, giving the *fatwa* a prescient nature. The third "fact" cites an established geopolitical issue, the Israel-Arab conflict and the plight of the Palestinians: however, it introduces new geopolitical actors into this situation. The sub-text is a claim that the leaders of Arab states have been unable to stop the Zionist policies of Israel, and so it is now up al-Qaeda to make a stand rather than politics as usual.

Following the components of a geopolitical code, the enemy identified by bin Laden's *fatwa* is clear; it is the United States and its "crusade" against Muslims, plus its regional ally Israel. Though the Saudi regime is strongly criticized, bin Laden "focuses on presenting the Saudi regime as a 'victim' whose dependent military and security relationship with the 'crusader forces' has led King Fahd to act subserviently to US interests and designs" (Ranstorp, 1998, p. 326). Rather than blaming particular groups or leaders, bin Laden was trying to identify a broad range of allies by emphasizing "unity." But these allies are nebulous, and certainly not a list of countries. Rather, the "we and you" that are called to arms suggests that it is interconnected individuals and groups who will enact the geopolitics of the *fatwa*.

The means of the geopolitical code were as simple as they were brutal. "We—with God's help—call on every Muslim who believes in God and wishes to be rewarded to comply with God's order to kill the Americans and plunder their money wherever and whenever they find it" (Ranstorp, 1998, p. 329). The means of maintaining allies rested in the interpretation of anti-US violence being the divine will of God; unity would come through "every Muslim" following God's will. The justification of such geopolitics was, for bin Laden, found within the Koran. The reference to divine will is the ultimate justification for action.

Bin Laden's *fatwa* was, of course, translated into horrific acts of terrorism that provoked the US invasion of Afghanistan, and the overthrow of the Taliban regime. More controversially, links between al-Qaeda and Saddam Hussein, partially retracted by President George W. Bush's administration, were used in the justification for the 2003 war on Iraq. The *fatwa* was a geopolitical code, still in operation, that challenges the world leadership of the United States. The "order" and "benevolent supremacy" of the United States is interpreted by bin Laden and his followers as a "crusade" for economic and religious reasons. For bin Laden, there is no leadership and common good within the United States' actions. Instead, the US's presence in the Middle East is seen as evidence of

> their eagerness to destroy Iraq, the strongest neighboring Arab state, and their endeavor to fragment all the states of the region such as Iraq, Saudi Arabia, Egypt, and Sudan into paper statelets and through their disunion and weakness to guarantee Israel's survival and the continuation of the brutal Crusade occupation of the peninsula.
>
> (Ranstorp, 1998, p. 329)

The allegation was that the United States was acting as an imperial power, dividing in order to conquer.

> **Activity**
>
> Am I correct in arguing that the concept of geopolitical codes is applicable to a non-state agent such as al-Qaeda?
>
> To justify your answer think about how the means of maintaining allies and engaging threats must differ for geopolitical agents other than countries, such as al-Qaeda.
>
> Must representation differ too?

The War on Terrorism as a geopolitical code

How did the world leader respond? Similar to NSC-68, the United States focused on two separate but related geopolitical agendas: protection of its sovereign territory and the construction of a global order. The defining document was the NSS of 2002, the foundation of what became known as the "Bush Doctrine." Within the framework of Modelski's model, it is the geopolitical code of the current world leader facing a challenge symptomatic of the beginning of the period of deconcentration.

The NSS is an annual exercise that updates the United States' geopolitical code. After the terrorist attacks of 9/11, the understandable focus was upon anti-American terrorism. By making the claim that the "struggle against global terrorism is different from any other war in history" (*NSS*, 5), the document was able to make the case that the established means to counter allies was ripe for change. The geopolitical threat identified by the NSS contained an apparent vagueness, but was able to become fixed on particular countries quite easily. The Strategy formalized the geopolitical code of the War on Terrorism, a war against "terrorists of a global reach" (*NSS*, 5). Simultaneously, this threat justified the global role of the world leader while also laying the foundation for action against specific countries: the "enemy is not a single political regime or person or religion or ideology. The enemy is terrorism" (*NSS*, 15). The clever use of "not a single" allows the code to be nebulously global and also, at times, geographically specific.

The vague and the specific were combined in the identification of the threat posed by "rogue states": countries that "brutalize their own people and squander national resources" (*NSS*, 9). Such acts are deemed a violation of the "basic principles" and goals of US world leadership. But rogue states are identified as a more specific threat too, being linked with the sponsorship of terrorism and the procurement of weapons of mass destruction. In this way, the notion of "rogue states" is able to give specific geographic definition, or targeting, to the general aims of world leadership (Klare, 1996).

With terrorism defined as the geopolitical threat facing the US, the "pre-emptive attack" was introduced as the legitimate means of countering the threat. The NSS evoked the United States' "right to self-defense by acting preemptively against such terrorists" (*NSS*, 6); simply to strike before "our enemies strike first" (*NSS*, 15).

In language that echoed NSC-68, the War on Terrorism was global in scope and historic in its intentions; "a global enterprise of uncertain duration" (*NSS*, opening statement). Matching our understanding of the role and tactics of a world leader, allies were to be maintained through "lasting institutions" (*NSS*, opening statement) that would provide the basis for "a truly global consensus about basic principles is slowly taking shape" (*NSS*, 26). The intention was to secure the continuation of US world leadership; "these are the practices that will sustain the supremacy of our common principles and keep open the path of progress" (*NSS*, 28).

Such is the language we would expect from a world leader, but Modelski's model suggests that the US will face an increasingly violent challenge to its authority. Not surprisingly, the NSS includes means other than institutions and "principles" to secure allies. Indeed, now "is the time to reaffirm the essential role of American military strength" (*NSS*, 29). But notably, the geography of this military strength was a global mission rather than the securing of the United States' borders: "The presence of American forces is one of the most profound symbols of the US commitment to allies and friends" (*NSS*, 29). Similar to NSC-68, the language of NSS balanced an identification of a threat to the US society and people, in terms of continued terrorist attacks, with a global commitment to promoting a particular vision of order; including economic relationships (Box 3.3). On the one hand, such order was deemed to be globally beneficial yet, on the other hand, it was "a distinctly American internationalism that reflects the union of our values and our national interests" (*NSS*, 1).

Box 3.3 The geopolitics of the "Washington Consensus"

The geopolitical agenda and power of the US is as economic as it is military. Its influence in the key global economic institutions of the World Bank, IMF, and World Trade Organization (WTO) is a reflection of its material interests and power to disseminate an ideological agenda. Indeed, since the 1990's the term "Washington Consensus" has developed as a summary of the economic policies that the US has pushed other countries to adopt, with much success. Under the umbrella of the term are policies of trade and investment liberalization, privatization, deregulation, fiscal and tax policy, and changes in the direction of public spending. Over time, those critical of such policies have also added issues of corporate governance, corruption, labor policy, and social safety nets into the argument.

("Washington Consensus" Global Trade Negotiations Home
Page, www.cid.harvard.edu/cidtrade/issues/washington.html.
April 2003—accessed 1/17/05)

In combination, these policies, whether they are seen positively or negatively, fall under the phrase "Washington Consensus": the economic side of the agenda of US world leadership.

In what way can the "Washington Consensus" be explained within Modelski's model?

The justification for the geopolitical code invoked language that was similar to that used in NSC-68: personal freedom was the goal, free-market economics was the means. The justification was targeted toward domestic and global audiences: "A strong world economy enhances our national security by advancing prosperity and freedom in the rest of the world" (*NSS*, 17). The Strategy promoted free trade as the economic vehicle, a policy that was portrayed as having benefits for everyone across the globe: "This is real freedom, the freedom for a person—or a nation—to make a living" (*NSS*, 18).

In a related statement, made at a time of confidence after the "victory" in Afghanistan that led to the removal of the Taliban regime by an American invasion as punishment for their support of al-Qaeda bases, President George W. Bush used his annual State of the Union speech to make focused geopolitical goals, within the framework of the War on Terrorism's global order. An "axis of evil," comprising Iran, Iraq, and North Korea was identified. The geopolitical threat posed by these states was, not just their alleged ties to terrorism, but also the identification of programs to build nuclear, chemical, and biological military capacity—weapons of mass destruction.

> Our second goal is to prevent regimes that sponsor terror from threatening our friends and allies with weapons of mass destruction. Some of these regimes have been pretty quiet since September the 11th. But we know their true nature. North Korea is a regime arming with missiles and weapons of mass destruction, while starving its citizens.
>
> Iran aggressively pursues these weapons and exports terror, while an unelected few repress the Iranian people's hope for freedom.
>
> Iraq continues to flaunt its hostility toward America and to support terror. The Iraqi regime has plotted to develop anthrax, and nerve gas, and nuclear weapons for over a decade. This is a regime that has already used poison gas to murder thousands of its own citizens—leaving the bodies of mothers huddled over their dead children. This is a regime that agreed to international inspections—then kicked out the inspectors. This is a regime that has something to hide from the civilized world.
>
> States like these, and their terrorist allies, constitute an axis of evil, arming to threaten the peace of the world. By seeking weapons of mass destruction, these regimes pose a grave and growing danger. They could provide these arms to terrorists, giving them the means to match their hatred. They could attack our allies or attempt to blackmail the United States. In any of these cases, the price of indifference would be catastrophic.
>
> We will work closely with our coalition to deny terrorists and their state sponsors the materials, technology, and expertise to make and deliver weapons of mass destruction. We will develop and deploy effective missile defenses to protect America and our allies from sudden attack. And all nations should know: America will do what is necessary to ensure our nation's security.
>
> We'll be deliberate, yet time is not on our side. I will not wait on events, while dangers gather. I will not stand by, as peril draws closer and closer. The United States of America will not permit the world's most dangerous regimes to threaten us with the world's most destructive weapons.
>
> (State of the Union Speech, 2002)

Figure 3.4 "Freedom Walk."

This passage contains the elements of a geopolitical code. The threat is identified; terrorism, state sponsors of terrorism, and weapons of mass destruction. A coalition of allies will be built and maintained. The means of obtaining these goals are, given that this was broad ranging political speech, understandably vague. However, the notion of pre-emptive strikes and the identification of specific countries, especially Iraq, were made clear. The justification of these actions was most direct: famine in North Korea, Iraqi "mothers huddled over their dead children," and "unelected" Iranian fundamentalist leaders were the opposites of the global order of prosperity, freedom, and civilization that the world leader had established as its agenda. Note the phrase "But we know their true nature"; a claim that the US has the ability to cast an all-knowing "God-like" eye across the globe. Public events, such as the "Freedom Walk," embed the geopolitical actions and the justifications for them in domestic places—making geopolitics a combination of "global" foreign policy and "local" everyday life (Figure 3.4).

Activity

To keep this chapter at a reasonable length I have had to limit the analysis of US foreign policy documents. The statements of Presidents Carter and Reagan may be especially useful for you to investigate, or even President Theodore Roosevelt at the beginning of the US's rise to power.

The contemporary codes of Turkey and China would also be intriguing to explore, given that the domestic politics of each one is fluid at a time when the country is increasing the geographic scope of its influence.

Summary and segue

Understanding the concept of geopolitical codes allows for an analysis of the multiple agendas that countries face and the diversity of policy options that are available to address them. Moreover, geopolitical codes are contested within countries as different political interests within a country seek different policies. Geopolitical agents do not have complete freedom in defining their code, the context of what other, perhaps more powerful, countries are doing must be taken into account. The dynamism of geopolitical codes is a result of the interaction, perhaps inseparability, of domestic politics and the changing global context. One way of defining the global context is through Modelski's model, though the concept of geopolitical code is still valid and useful without resort to the model of world leadership. In the next chapter, we will concentrate on the fifth element of geopolitical codes: the way they are represented to gain public support.

Having read this chapter you will be able to:

- define geopolitical codes;
- interpret government foreign policy statements as the manifestation of geopolitical codes;

- consider the actions of geopolitical agents other than countries as the manifestations of their geopolitical codes;
- consider how the content of geopolitical codes is influenced by the global geopolitical structure;
- consider how the global geopolitical structure is changed by the actions outlined in geopolitical codes.

Further reading

Flint, C. and Falah, G.-W. (2004) "How the United States Justified its War on Terrorism: Prime Morality and the Construction of a 'Just War'," *Third World Quarterly* 25, pp. 1379–99.

A discussion of how the United States, as world leader, has different needs, and uses different language, in justifying its geopolitical code compared to other countries.

Halliday, F. (1983) *The Making of the Second Cold War*, London: Verso.

An excellent discussion of the actions of the United States and Soviet Union in the Third World that provides background for the discussions of US geopolitical codes.

Klare, M.T. (1996) *Rogue States and Nuclear Outlaws: America's Search for a New Foreign Policy*, New York: Hill and Wang.

Provides background to current geopolitical pronouncements regarding "rogue states" and the "axis of evil."

Ranstorp, M. (1998) "Interpreting the Broader Context and Meaning of Bin Laden's *Fatwa*," *Studies in Conflict and Terrorism* 21, pp. 321–30.

As the title indicates, this article provides background and context for bin Laden's initial call for a conflict with the United States.

REPRESENTATIONS OF GEOPOLITICAL CODES

In this chapter we will:

- introduce the cultural aspect of geopolitical codes;
- focus on the ways in which geopolitical codes are justified;
- identify the linkage between popular culture and foreign policy;
- introduce the concept of Orientalism;
- discuss how popular culture helped the public interpret the content of NSC-68;
- discuss how Saddam Hussein used a combination of Arab nationalism and Muslim belief to justify his 1990 invasion of Kuwait;
- discuss how the administration of President George W. Bush represented its War on Terrorism.

The previous chapter concentrated upon how geopolitical codes were formulated within the structure of global geopolitics. An essential dimension of a geopolitical code is the way that a country's decisions and actions are justified. A convincing case for why a country is a "threat" or not, and what should be done about it, must always be made not only to a country's own citizens, but also to the international community. This chapter will explore how violent acts of geopolitics (the prosecution of wars) are portrayed as the defense of a country's material interests plus its values. Detailed discussion of the rhetoric of Saddam Hussein's representation of the Gulf War of 1991 illustrates how domestic support for a war was fostered. The final section of the chapter describes President George W. Bush's justifications for the 2003 invasion of Iraq—an argument that was addressed equally to domestic and global audiences.

War! What is it good for . . .?

On the surface, the "Soccer War" between El Salvador and Honduras provides an illustration of how petty national concerns and hatreds can explode into warfare. The value of "national pride" was marshaled to provoke and justify a war. However, just focusing on national differences, in this case, is a shallow and incomplete understanding, as we shall see. In 1969, El Salvador and Honduras played two games of football ("soccer") in the qualifying stages for the 1970 World Cup finals (see Kapuscinski, 1992, pp. 157–84, for a full narrative of this conflict). The first game, in Honduras, resulted in a one-nil victory for the home side. Back in El Salvador, eighteen-year-old Amelia Bolanios committed suicide in light of the national shame. Her funeral was a national event, the procession led by the president of El Salvador and his ministers. The return match in El Salvador was played in an extremely hostile atmosphere; El Salvador won three-nil. The Honduran team retreated to the airport under armed guard, their fans were left to their own devices and two were killed as they fled to the El Salvador–Honduras border. The border was closed in a matter of hours. The Honduran bombing of El Salvador and military invasion followed shortly afterward. The war lasted 100 hours, 6,000 people were killed and 12,000 wounded; the destruction of villages, homes and fields displaced approximately 50,000 people.

But are nationalist passions sparked by football matches enough to initiate the horrors of war? Underlying the tension between El Salvador and Honduras, a tension that easily aroused national hatred as footballs landed in goal nets, was a struggle for land and human dignity that crossed an international border (Kapuscinski, 1992, pp. 157–84). The land of tiny El Salvador, with a very high population density, was owned by just 14 families. In a desperate attempt to obtain land, about 300,000 El Salvadorans had emigrated, illegally, across the border and established villages. The Honduran peasants also wanted land reform, but, and backed by the US, the Honduran government avoided redistributing land owned by its own rich families and the dominant United Fruit Company. To avoid an internal political struggle, the Honduran government proposed to redistribute the land that the El Salvadorans had settled. The prospect of forced repatriation from Honduras not only unsettled the migrants, but also rattled the government of El Salvador who faced the prospect of a peasant revolt.

Landlessness, monopoly, human dignity, fear of popular rebellion: these mutual "domestic" issues were intertwined across the porous Honduras–El Salvador border. The government's decision to go to war was made within a context of class inequality and the inequities of land ownership. National humiliation on the football field was merely the fuse that lit the political tinderbox. International war was deemed a more obvious solution than altering the domestic status quo.

A more recent consideration of the role of material "needs" and ideological hype in oiling the movement toward war was evident in *Fahrenheit 9/11* and its portrayal of the geopolitical doctrine of President George W. Bush. According to filmmaker Michael Moore, the war upon Iraq was a matter of capitalist greed. The maintenance of personal fortunes built upon the global need for oil. Unsurprisingly, the reasons President Bush gave for going to war are different. They did not rest upon the material needs, and financial gains, of extracting and selling oil. Rather they rested in the realm of ideals. According to this interpretation, the war was fought in the name of "freedom" and securing the privileges of the citizens of the United States in the face of terrorist threat. Also, the ideals of "liberating" the Iraqi people from tyrannical rule and bestowing them with democratic self-rule were cited. As in the Soccer War, contemporary discussions of war juxtapose two, equally political, interpretations of the causes of war: material gain and values.

Answering the question headlining this section is clearly beyond my capabilities. But let me try and provoke an initial approach to the question in a way that provides some insights into particular conflicts while also placing war within our broader discussion of geopolitics. Our discussion will focus on two different reasons for fighting wars, specifically the reasons governments use to justify their involvement in conflicts: material interests and values. These two reasons should not be seen as competing or mutually exclusive. Instead, they are presented as the two most common themes used to justify participation in warfare.

A prime philosopher of the material motivations for war, V.I. Lenin (1939) was writing at the time of the global war identified by Modelski as the end of British world leadership. For Lenin, the leader of the Bolshevik Revolution in 1917 and first premier of the Soviet Union, the upcoming wars were materialist in nature, an expression of the imperialism of the rich powers needing new markets and sources of raw materials to feed the banks and finance groups within their borders. For Lenin, the two world wars were the bloody component of the continuous struggle for profits. The Soccer War and the war on Iraq could be interpreted the same way.

Alternatively, sociologist Pitirim Sorokin (1937) argued that war is fought over competing values. The national humility fatally felt by Amelia Bolanios in El Salvador in 1969 was a sign of the power of values in warfare. The impassioned speeches of President George W. Bush and Prime Minister Tony Blair regarding Iraqi liberation frame the invasion of Iraq, and other episodes of the War on Terrorism, as a conflict over values: values that are deemed to justify loss of life among coalition forces, humanitarian workers, insurgents, and Iraqi citizens.

Rather than attempting to portray, and resolve, a simple debate between a "materialist" and a "values" perspective on war, the aim of this section is to initiate an exploration of the different geographies of representation that result from the material and value interpretations of war. Representations of war that were based upon material concerns or "interests" are territorially based, often reflecting concerns over control of territory or boundary location in order to access key resources. On the other hand, representations of war that resort to ideals are less bound to specific pieces of territory, and tend to speak to visions of what is best or, "common sense", for humanity.

Activity

For any foreign policy event of your choice (a war, the imposition of sanctions, the establishment of alliances, etc.) look at policy documents, speeches, or media commentaries that portray the policy and evaluate the degree to which justification was made through material interests or values.

Are the relationships between material justifications/territoriality and value-based justifications/extra-territoriality I posit evident?

Cultured war

Dulce et Decorum Est pro Patria Mori (It is sweet and proper (or fitting) to die for one's country) (Box 4.1). It is only of late that Hollywood has begun to portray the horror, pain, loneliness, and indignity of dying in war. Movies such as *Platoon* told a story of the Vietnam War. Steven Spielberg's *Saving Private Ryan* was technically brilliant in showing the terror, confusion, and slaughter of the Normandy landings of World War II, but its main purpose was an act of remembrance and national thanks for the

---

Box 4.1 *Dulce et Decorum Est*

Bent double, like old beggars under sacks,
Knock-kneed, coughing like hags, we cursed through sludge,
Till on the haunting flares we turned our backs
And towards our distant rest began to trudge.
Men marched asleep. Many had lost their boots
But limped on, blood-shod. All went lame; all blind;
Drunk with fatigue; deaf even to the hoots
Of tired, outstripped Five-Nines that dropped behind.

Gas! Gas! Quick, boys!—An ecstasy of fumbling,
Fitting the clumsy helmets just in time;
But someone still was yelling out and stumbling
And flound'ring like a man in fire or lime . . .
Dim, through the misty panes and thick green light,
As under a green sea, I saw him drowning.

In all my dreams, before my helpless sight,
He plunges at me, guttering, choking, drowning.

If in some smothering dreams you too could pace
Behind the wagon that we flung him in,
And watch the white eyes writhing in his face,
His hanging face, like a devil's sick of sin;
If you could hear, at every jolt, the blood
Come gargling from the froth-corrupted lungs,
Obscene as cancer, bitter as the cud
Of vile, incurable sores on innocent tongues,—
My friend, you would not tell with such high zest
To children ardent for some desperate glory,
The old Lie: Dulce et decorum est
Pro patria mori.

(Wilfred Owen, 1917)

World War II generation: supported by the book and film *Band of Brothers*. None of these efforts come close to the cynicism of Wilfred Owen's poem; for Owen juxtaposes the brutality of individual death, with the romantic mythology of nationalism. As the soldier Owen describes is feeling life slip away as his lungs are being corroded by gas, is he really going to reflect on the "sweetness" of his duty to give his life for his country? In actuality, the common cry of the dying soldier, usually a young man, is for their mother (Fussell, 1989).

Yet, at the beginning of World War I, millions of people across the European continent and within Britain greeted the outbreak of war with unbridled joy (Eksteins, 1989; Tuchman, 1962). People lined up to join their respective military; it seemed like a great thing to be going off to war. Owen's cynicism came later, and was a product of experience at the front, and a reaction to what he saw as the inhumanity of nationalism driving young men to their deaths.

World War I is widely seen as the epitome of the modern war (Eksteins, 1989), but it also ushered in the rise of fascism, especially Adolf Hitler and the Nazi party. We may all be familiar with the term Nazi and Nazism, but it is important to reflect on the meaning of the name. "Nazi" stems from the full title of Hitler's party National Socialist German Workers Party. The "national" and the "socialist," or the emotive and the material combined powerfully in Hitler's ideology to give one of the clearest, and most reviled, expressions of nationalism in history. Nationalism is the belief in a common culture, or people, and its connection to a particular country. The term will be discussed in greater depth in the following chapter. Hitler's rhetorical strength lay in his ability to link material grievance with an ideologically based future within a portrayal of the German national past and future glory. Though Nazism is an extreme, the extreme simply serves to illuminate what is common in contemporary geopolitics. Representations of war and other forms of geopolitics are usually based within an understanding of an individual's membership in a national group, which has its particular values, traditions, and history (Figure 4.1).

Geopolitical actions are given meaning in order to justify their prosecution. Geopolitics is, then, a cultural as well as a political phenomenon, and usually a national one. Culture normalizes the continuous prosecution of geopolitics across the globe. More specifically, it paints "our boys" (and to a much lesser extent "our girls") as heroes fighting a valiant and necessary fight, while portraying the enemy, or "them," as evil and villainous (Fussell, 1990; Hedges, 2003). Increasingly, "they" are made invisible— deaths that we need not worry about as we prosecute war (Gregory, 2004).

Me, a geopolitician? Laughable!

First, let us explore what we know without knowing, or at least without thinking or questioning. **DO NOT LOOK AT BOX 4.2 until you have read the next few lines.** First, make a column of numbers from 1 to 12. Second, get ready to look at the list of countries in Box 4.2. Don't look yet. I want you to read the name of each country in turn and write the first word that comes into your head—no matter what it is. The key to this exercise is not to think too deeply, and not to worry about what you are writing. Tip: don't write the name of the country, just move through the list quickly. Go!

Figure 4.1 World War II memorial, Stavropol, Russia.

Once you have written the list you can consider the following questions:

1   What are the sources of the images or ideas behind the word you wrote?
    Think of movies, news reports, books, lectures, magazines, songs that have created
    a picture of a country for you—one positive and one negative.
2   Do these images reflect particular groups in society? In other words, do you think
    the image comes from a male or female perspective, a white or other racial posi-
    tion, an elite or non-elite group?
3   What are the implications of these images to the foreign policy of your country?
    In other words, do these particular images and the response they generated in your
    mind facilitate particular policies?
4   Which terms or words that you came up with lead to justification of foreign policies
    that were either violent or required no action?

This simple exercise is trying to suggest that we all carry around "knowledge" of
countries that we probably know very little about. This "knowledge" is gained from the
most dubious sources, primarily Hollywood movies and television shows, and complemented by songs, jokes, and comedy routines, etc. It is nothing new. As a boy of about
seven or eight years old, I can remember my grandfather playing a recording to me by
a comedy duo, Flanders and Swan. One of their songs was called something like "The
English are best, I wouldn't give tuppence for all of the rest." It was a list of all the

---

Box 4.2 Geopolitical word association

1   United States
2   North Korea
3   France
4   Columbia
5   Afghanistan
6   China
7   Turkey
8   Iraq
9   Japan
10   Democratic Republic of the Congo
11   Pakistan
12   Great Britain

---

peculiar faults and traits that are supposedly possessed by different national groups, and in the process expunges any negative characteristics from each and every English person. This may seem harmless, but it is powerful because it is pervasive and everyday. Listening to the record of an evening was "family fun," that just happened to instill a belief that my country was obviously superior to any other. Such "humour" was the basis for a geopolitical understanding of Britain's "right" to tell other countries, using force if necessary, what to do.

My whole generation grew up in England on a steady diet of "Irish jokes": continually painting an image of all Irish people as hopelessly stupid. How could I then, as I grew older, begin to think there was a historical basis for Irish nationalism? A deeper understanding of this conflict, and others, had to be actively sought by myself despite the obstacles of the "common knowledge" provided by mainstream media sources and cultural attitudes. My knowledge of the Irish had been created by the English media and the telling of Irish jokes at the back of the bus; what else did I need to know? The playground, the bus stop, and the couch in front of the TV were (and still are) very important arenas for an understanding of geopolitics. The basis for these images was not just schoolyard jokes passed down from older to younger siblings, but also the result of cultural products such as movies, books, magazines, and songs.

In the Gramscian sense of power, we carry with us "knowledge" of the world that is often of the must dubious and partial nature, but the knowledge is powerful nonetheless. Its power comes from it being taken for granted as "common sense," on the one hand, and in the way that knowledge is the foundation for the "ideals" used to justify geopolitical actions. For example, if whole swathes of the world are deemed "anarchic" then policies combining non-involvement in some cases (such as Rwanda) or military intervention (such as Afghanistan) in other cases may be implemented with little need to explain or defend them. Of course, production of the cultural common sense underlying foreign policy cannot be left to the imagination of playground humorists; the media industry is heavily implicated.

"Freedom," "slavery," Hollywood, and the *Reader's Digest*

In the previous chapter we discussed NSC-68 as a defining document of the US's geopolitical code. NSC-68 was not America's bedtime reading in 1950. The global geopolitical code of the United States had to be disseminated to the public in more appealing media. With regard to the role of the United States in one key region, "the Middle East was not immediately available as an American interest; instead, it had to be made 'interesting'" (McAlister, 2001, p. 2). Films, museum exhibits, and television news were all brought to bear, and still are, upon American and global audiences, to represent America's world leadership in ways that would justify its geopolitical code.

The assumption of the status of world leader ushered in a slew of geopolitical responsibilities. The global diplomatic and political presence of the US was heightened, in terms of its size and the accompanying drama, with the development of the Soviet Union's challenge to its global role and the ensuing Cold War. In the Middle East, crises over the establishment of Israel and the displacement of the Palestinians, and the competition with the Soviet Union over the establishment of "friendly regimes" to help secure access to oil required a growing US presence in the region. How was this to be portrayed to the American people? The first step was to create a particular "knowledge" of the Arab world: a "knowledge" that would act as the foundation for seeing the necessity and value of US influence in the region.

A spate of biblical and Roman "epics," such as *The Ten Commandments* and *Quo Vadis*, brought visions of the "Middle East" and classical "values" to the American, and global, public: but it was a very particular vision. The Middle East was seen as "the Holy Land," and specifically the themes of slavery and gender relations were emphasized, and reinforced in the Roman stories. McAlister's (2001) study of these movies shows that Hollywood superstar actors and actresses, often in ridiculous costumes, interacted in a Hollywood-style romance with an obvious sub-text (Figure 4.2). The woman was a victim of slavery; what else could the beautiful object do given the despotic, barbaric, and cruel society she was shown to be living in? But, do not fear. She was "liberated"—clearly a word with geopolitical overtones—from this role by the actions of the male hero, with the American accent. After "liberation" what did the heroine do? Form a radical feminist alternative society? Not surprisingly, no. She willingly entered a subservient role as the wife of the male hero—in other words after despotism the choice is made to be dominated. "Liberation" is too much, perhaps the movie suggests it is a role a woman is unsuitable for; she needs to be controlled or aided in some way.

> Working through a gendered logic that figured "slavery" in sexual terms—as a problem for (white) women in relation to despotic men—the films offered right-ordered marriage and the "freely chosen subordination" of women as the solution. They then cast that subordination as a model for the relationship between the United States and the decolonizing nations of the Middle East, constructing US power as a "'benevolent supremacy' that would replace older models of direct colonial rule".
>
> (McAlister, 2001, p. 40)

Figure 4.2 Hollywood biblical epic.

The "slavery" of the movies had two geopolitical referents: the threat of Soviet expansion and the yoke of Communism, on the one hand, and the historic chains of European colonialism on the other. The films carried a message that the order imposed by the United States' world leadership would break these chains while preventing the imposition of another yoke. The geopolitical code of NSC-68 sought supremacy or leadership for the US, but the goal was benevolent, global order through the operation of free institutions.

The message of these movies is part of the cultural "knowledge" that facilitated the actions of the United States in the Middle East. Action was deemed necessary to "save" innocent and helpless people from despotic and barbaric circumstances. However, "liberation" alone was not enough. "Protection" had to be given too, in a relationship that is on the one hand dominant, but at the same time acceptable and moral because it is desired by the weaker party. NSC-68's appeal to free people everywhere is then not only a matter of "liberation" or "salvation" from despotism, but also a justification for geopolitical influence in societies portrayed as too weak or "young" to be able to stand on their own two feet.

Throughout the twentieth century, the *Reader's Digest* provided a source of geopolitical commentary for the American public. A widely read magazine with a broad and loyal audience, the *Reader's Digest* has captured the consistent and changing geopolitical messages that were aimed to help its readers understand, in order to support, the global geopolitical role the US had assumed (see Table 4.1) (Sharp, 2000b).

Table 4.1 Reader's Digest *articles identifying threats to the American way of life*

	1986–8	1989–91	1992–4
Russia and Communism	48	50	34
Terrorism	6	8	1
Drugs	13	19	5
Japan/economic	7	9	10
Domestic danger	27	51	76
American values	13	11	25

Source: Slightly modified from Sharp (2000, p. 339).

Note: By "domestic danger" the *Reader's Digest* means the bureaucratic influence of big government, as well as the spread of "political correctness."

A consistent theme throughout the century was the comparison of the US to classical societies who had given so much to humanity through the diffusion of civilization. In 1922 the *Reader's Digest* stated:

> The art that was Greece and the legal temperament that was Rome reflect the idealism of great peoples who had something within their national souls which became the common heritage of humanity. This is the supreme test of a truly great people. . . . No fear need be felt that the historians of the future will pronounce national humanitarianism—the will to disinterested human service—the original national contribution of the United States to the higher idealism of the world.
>
> (Sharp, 2000a, p. 75)

In this quote, the global mission of the United States is portrayed as a benevolent act, disseminating a humanitarianism that will benefit all, similar to a mythic interpretation of the Romans, the provider of the rule of law, and not the efficient and technologically superior invaders of other countries. Note too that the "duty" and the ability of the US to conduct this global mission rests not in government and military power but in national characteristics. In other words, the individual American has a global role to play, not just the NSC, the Pentagon, and the president. *Reader's Digest* paid greatest attention to the Soviet Union in the 1950s and 1960s. In the 1980s and 1990s new threats were identified, including growth of "big" government and "political correctness" (Sharp, 2000b, p. 339).

Two related threats to "America" were identified by the *Reader's Digest*. On the one hand was the concern of degeneracy, or the fact that the US's power could wane, similar to Rome's classical fate (Sharp, 2000a, p. 75). On the other hand, was Communism, portrayed as a threat to humanity as well as the culture and moral fabric of America, as the magazine argued in 1935 and again in 1950 and 1948:

> The American dream of Poor Boy makes Good leads even the most underpaid drudge to consider himself a potential millionaire. This makes it hard to arouse him to a Marxian consciousness.
>
> (Sharp, 2000a, p. 77)

> The unsuspecting American imagines that we are safe from socialism because
> he knows the people will never vote for it. But socialism can be put over by a
> small minority.
>
> (Sharp, 2000a, p. 89)

> The power of the Soviet Union, and particularly the Soviet Communist Party,
> is due to the fact that, while in a sense the Soviet state has moved into a power
> vacuum in Europe and Asia, the Soviet Communist Party has moved into a
> moral vacuum in the world.
>
> (Sharp, 2000a, p. 89)

Sharp shows how the *Reader's Digest* was able to connect the US's global geopolitical agenda to the concerns and actions of the individual citizen. Wars in Korea and tensions in Berlin, for example, were represented as the necessary outcomes of a mission bestowed upon the US because of its national character. The Cold War was portrayed as a conflict over values: US humanitarianism versus Soviet totalitarianism. Moreover, the "battle" was not geographically distant. The *Reader's Digest* made linkages across spaces and down geopolitical scales, so that an immediate threat to the fabric of society was constructed. The result was that geopolitical agency was essential for all Americans, whether they were fighting Communism in a foreign land or working as a farmer, teacher, shop assistant, etc. in Nebraska. The *Reader's Digest* made this clear in 1952:

> This is where *you* come in. No-one is too small or insignificant, too young or
> too old, to be shackled and regimented, or pauperized and destroyed. . . . By
> its all-encompassing timetable sooner or later [the "communist masterplot"]
> has to reach you.
>
> (Sharp, 2000a, p. 93)

Individual morality as a form of everyday geopolitical agency remained an important theme for the *Reader's Digest* after the Cold War. Increasingly, "domestic danger" became the geopolitical focus of the magazine, as the threat of Russia and Communism declined through the 1990s (Sharp, 2000a, p. 151). Continuing a theme that began with the New Deal program, in 1994 the *Digest* noted that "[t]he barbarian's are not at the gates. They are inside" (Sharp, 2000a, p. 152): continuing the magazine's reference back to the classical ages of Rome and Greece, made accessible by the Hollywood biblical epics discussed earlier (McAlister, 2001). Who were the barbarians? Surprisingly, it was the American government itself, or more specifically the portrayal of an increasingly powerful central government and, in the argument of the *Digest*, a consequent culture of dependency upon government services. In themes that loyal readers would be able to trace back to early discussions of Communism, government intervention (or degrees of Communism) created the potential of moral decay.

In sum, the *Reader's Digest* offered a commentary on US global geopolitics that assigned an exceptional nature to the country that demanded a global presence. Participation in wars across the globe, and other forms of political intervention and presence, were related to the everyday life experiences of the readers by a representation of the American nation as one embracing particular moral standards and norms. In this manner geopolitical agency was portrayed as a daily necessity for the individual; as Communism and big government threatened American communities and, by extension, the US's ability to conduct its humanitarian mission for the whole world.

Activity

The *Reader's Digest* and the Hollywood biblical epics are an example of both the Gramscian and feminist definitions of power.

- Look at Figure 4.3 and identify the gender roles that are portrayed.

- Discuss how the gender roles build upon to our taken for granted assumptions about how foreign policy is conducted and by whom.

Our embassies abroad continue to hire local employees who are some-times as adept at pampering our foreign-service people as they are at espionage

Why Our Embassies Are Nests for Spies

Condensed from
WASHINGTON MONTHLY

PRISCILLA WITT

AMERICANS WHO take assign-ments at U.S. embassies around the world have al-ways been warned about foreign agents who—through seduction, bribery, bugging or other means—will attempt to recruit them to ob-tain classified information. In the wake of the recent Marine spy scan-dal in Moscow, the United States has witnessed just how real the dangers are.

Even more startling, we have seen that these spies needn't hang out in smoky clip joints, waiting to ensnare unsuspecting embassy em-ployees. Instead, today's foreign agents can sometimes be found in-side the protective walls of Ameri-can embassies, as paid employees of the U.S. government.

Our Moscow embassy, which once had 220 Soviets on its payroll,

Figure 4.3 "Nests for spies."

Orientalism: the foundation of the geopolitical mindset

Representations of geopolitics, with clear messages regarding personal and national behavior, are embedded within a whole host of media that "entertain" and "inform" us, without claiming to be overtly political. They are in movies and books etc., but the presence of these representations in readily accessible media is the product of much deeper cultural structures that go under the title of Orientalism.

Edward Said (1979) was the driving force behind the concept of Orientalism, by which he meant the institutionalized portrayal of non-Western cultures as "uncivilized," "backward," "child-like" even "barbaric" and "primitive" in such a manner that it pervaded government, academic, and popular culture circles. Said was a professor of English Literature and analyzed novels, especially by English authors, in the nineteenth century. However, his work is still relevant today, and is the basis for many academic works on how "knowledge" of other cultures is created and disseminated. Furthermore, the point of Orientalism is that such "knowledge" of, say, Arabs, or Muslims, or Africans is a form of power. There is power in the ability of Western countries to create particular understandings of the rest of the world, or classify weaker countries and their inhabitants. For example, Western media portrayals of African countries are pervasive, African representations of Europe and the United States are not. Such knowledge becomes unquestioned because it is seen everywhere. Second, the authority of the knowledge, given that it is largely unquestioned or countered by alternative images, allows for, or demands, particular foreign policy stances toward particular countries. Orientalism is the foundation of the responses to the geopolitical word game we played earlier (Box 4.2). North Korea *is* nuclear weapons, for example.

But Said did not only point out that the West portrayed non-Westerners as barbarians to justify their colonization. There is a double-sided nature to the process too. By portraying non-Westerners as "backward" and "uncivilized," etc. Western countries and their geopolitical practices were painted, for self-consumption, as the exact opposite: "modern," "the bearers of civilization," etc., and hence the "natural" rulers of the globe. This self-portrayal of the West was not done just to make people feel good about themselves: the extremely brutal acts of conquest and oppression that were necessary for the West to establish its imperial rule over the world could then be seen as the required, if unfortunate, acts needed to "discipline" or "civilize" the "natives." If the competitive colonization of Asia was known as the "Great Game" in a reference to the sports-field escapades of the British ruling class, then the household belief that "to spare the rod is to spoil the child" was also transferred to the global scale—in the belief that "natives" only understood discipline. Orientalism did not die with the end of formal empire. In fact, it has been noted that the portrayal of vast numbers of human beings as "savages" and "barbarians" has been in resurgence in the wake of the terrorist attacks of September 11, 2001 (Gregory, 2004).

The *profession* of Orientalism, as Edward Said called it, continues today. Academics at respected universities write books, newspaper columns, and make television appearances that combine to tell us the world out there is full of savage and irrational people just waiting to inflict pain and suffering on the innocent West. Kaplan's (1994) "Coming

Anarchy" is a good example, as is, with specific focus on Arab countries and Muslims, the work of Bernard Lewis (2002). Increasingly, the target of Orientalism has become Islam—a topic readily adopted by Fox News in the US, and British talk-show host and now politician Robert Kilroy-Silk, for example. These everyday acts of portraying a dangerous world needing US policing, with the help of other Western countries and especially Britain, is established on the nearly 200 years worth of cultural products first analyzed by Said. The contemporary catalyst was Samuel Huntington's (1993) "The Clash of Civilizations": epitomized by its classification of the world into eight "civilizations"—the most problematic one being Islam with its "bloody innards and bloodier boundaries." Empirical analysis does not support Huntington's bold claims —in statistical analyses of conflicts across the world, connections to Islam does not increase the likelihood of war (Chiozza, 2002). But who reads the academic journals? The talking heads and op-ed pieces are the "high-brow" contributors to "common-sense" and "Indiana Jones" the low-brow.

Activity

Consider the movie releases in your hometown over the past, say, six weeks.

- Who were the enemies or "baddies" portrayed in the movies?
- Do they represent, either overtly or subtly, real world countries or other geo-political agents?
- Who do the "goodies" represent?
- What were the nationalities of the actors who played the "goodies" and the "baddies"?
- Consider the gender roles in the movies; can you trace geopolitical messages akin to the interpretation of the biblical epics we discussed earlier?

Scholars were quick to point out the cultural misrepresentations in Huntington's work, but it still, along with the work of Robert Kaplan, sowed the seeds of a post-Cold War understanding that the world was "chaotic," "messy," and "dangerous" and hence needed "order" and "stability" (Dalby, 2003; Flint, 2001). Perhaps more insidious is the contemporary Orientalist practice of making whole populations invisible. The use of biblical epics was a cultural representation that demanded geopolitical involvement to "save" particular countries (McAlister, 2001). In the war on Iraq, the language has changed significantly, according to Gregory (2004). Iraqi people are dehumanized—either by making them invisible by just not mentioning them, or portraying them as "savages," beyond our civilized codes and not deserving of political or economic support.

The new media representations of satellite images, and computer simulations allow the Western viewer to be a virtual participant in the War on Terrorism. Geographer

Derek Gregory (2004, pp. 197–214) talks of this development at length; noting the interactive websites of *USA Today* and the *Washington Post* that allowed you to point and click over Baghdad, retrieve "details" of the targets, and keep track of the war by seeing photos before and after the bombs were dropped. At the same time, the images were almost completely empty of pictures of human suffering and carnage. Someone in Birmingham, Alabama or Birmingham, UK could "repeat the military reduction of the city to a series of targets, and so become complicit in its destruction—and yet at the same time . . . refuse the intimacy of corporeal engagement" (Gregory, 2004, p. 205). Geopolitics has become just another computer game of killing the bad guys, only in this case the victims are not just computer-animated figures they are absent. The essential point that Gregory is making by focusing on websites and computer games is that Orientalist representations are now something that the general public actively participate in and help create rather than being "fed."

At the same time, death and suffering is officially absent, in a breach from historic military practice the deaths of enemy combatants and non-combatants were not counted in the invasion of Iraq and the subsequent insurgency. Gregory's use of blunt official statements is most effective: "we do not look at combat as a scorecard" and "[w]e are not going to ask battlefield commanders to make specific reports on battlefield casualties" (Gregory, 2004, p. 207). From the Western perspective, contemporary war can be a computer game, just as long as you do not keep track of the human consequences: maybe it is the only computerized conflict available that does *not* allow you to count points!

However, in a time of electronic and globalized media, alternative visions are available. The Al-Jazeera satellite television was broadcasting images of carnage in Iraq across the Arab world; broadcasting pictures described by its editorial staff as "the horror of the bombing campaign, the blown-out brains, the blood-spattered pavements, the screaming infants and the corpses" (Gregory, 2004, p. 208). In a change from the nineteenth- and early twentieth-century Orientalist situation described by Said, the technology to broadcast the story of the victims is now possible. However, the legacy of Orientalism lies not just in the ability to broadcast, but what gets seen and how it is interpreted. Here the Western powers still have some advantage. Images from sources other than CNN, BBC, or ITN, etc., are easily dismissed as "cinematic agitprop," or stories reported "from the enemy side."

The geopolitical code of the world leader has always required a cultural complement to give it meaning that encourages popular support. As technology has changed then so has the role of the public: becoming "embedded" in such a way that the creation and consumption of geopolitical information becomes blurred (der Derian, 2001). But also, the ability for the "colonized" to speak back and give their own version has been enhanced too. The Internet is a means for people across the globe to be given the perspective of what it feels like to be "liberated" (see Box 4.3). Within Modelski's (1987) model of world leadership, the cultural message of "leadership" is still touted, but there are alternative interpretations too, making some attempt to undermine a singular view of the world as needing and requesting a benevolent US presence.

Box 4.3 Baghdad blog

For me, April 9 was a blur of faces distorted with fear, horror and tears. All over Baghdad you could hear shelling, explosions, clashes, fighter planes, the dreaded Apaches and the horrifying tanks tearing down the streets and highways. Whether you loved Saddam or hated him, Baghdad tore you to pieces. Baghdad was burning. Baghdad was exploding. . . . Baghdad was falling . . . it was a nightmare beyond anyone's power to describe. Baghdad was up in smoke that day, explosions everywhere, American troops crawling all over the city, fires, looting, fighting and killing. Civilians were being evacuated from one area to another, houses were being shot at by tanks, cars were being burned by Apache helicopters. . . . Baghdad was full of death and destruction on April 9. Seeing tanks in your city, under any circumstances, is perturbing. Seeing foreign tanks in your capital is devastating.

April 9, 2003 was the day Baghdad was declared to be under the control of American troops. The quote is from the weblog of Riverbend, an Iraqi woman, quoted in Gregory, 2004, p. 213.

Case study 4.1: Saddam Hussein's use of Arab nationalism and Islam to justify in the 1991 Gulf War

The 2003/4 war on Iraq was initiated by the United States within the parameters of the "War on Terrorism," a key component of its geopolitical code. This war must also be seen as a continuation of an earlier conflict, the Gulf War of 1991 when the US responded to Iraq's 1990 invasion of Kuwait in order to maintain the political status quo in the Middle East and counter Saddam Hussein's attempts to enhance Iraq's power. Looking at Saddam Hussein's use of political rhetoric during the Gulf War provides insights into key ingredients of Arab geopolitical codes as well as the tools used in an attempt to justify the invasion of Kuwait to Iraqis and the wider Arab population.

Underlying Hussein's actions and rhetoric were the politics of Arab nationalism (Figure 4.4). Arab nationalism can be viewed in two ways. First, as exemplified by Gamal Nasser (president of Egypt from 1954–70), Arab nationalism was a political agenda focusing upon Arab unity, a common nation of Arabs that would come together to resist external control by France, Great Britain, and the US, fight the state of Israel, and provide peace and prosperity for the Arab world. As part of this process, Syria and Egypt were united for a brief political moment between 1958 and 1961 when they formed the United Arab Republic. The politics of this expression of Arab nationalism was modern and secular, in opposition to traditional Islamic conservatives (see Khashan, 2000 and Mansfield, 1992 for more on Arab nationalism).

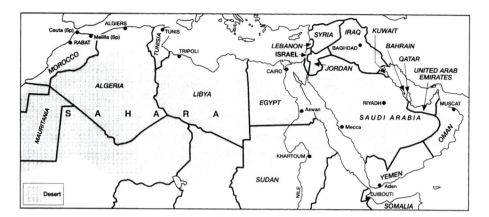

Figure 4.4 The Arab world.

However, the lifespan of the United Arab Republic was brief. It fell apart because of another expression of Arab nationalism, that each Arab state should be an independent sovereign state. The geographic scale of Arab nationalism has, in practice, been centered upon separate national interests. However, as we will see in the rhetorical politics of the 1991 Gulf War, the idea of the unified Arab world was still a backdrop to the politics of the Middle East.

The catalyst for Arab nationalism, in the sense of Arab unity, was the establishment of the state of Israel in 1948, and the consequent wars with Arab states (we will discuss the Arab-Israel conflict in more detail in Chapter 6). Israeli victories in the consequent wars of 1967 and 1973 led to a feeling of Arab humility, and further derailed the unity of Arab nationalism as individual countries made peace with Israel and also allied themselves with the United States, while others proclaimed their anti-Israeli and anti-colonial credentials in the name of Arab nationalism.

Two countries that remained hostile to Israel and the United States were Syria and Iraq. These two countries shared the Baa'thist political ideology, a combination of nationalism and socialism that had its philosophical roots in European left-wing politics. In that sense, it was an imported political ideology that aimed for the secular modernization of the Arab world in order to fight Israel and resist outside "colonial" influence, especially the presence of the United States. As we discuss the geopolitical code of Saddam Hussein, it is important to note that at the outset the Baa'thist philosophy was opposed to conservative fundamentalist Islamic political ideologies, but emphasized an Arab path to modern society.

Before the conflict with Kuwait, Iraq's geopolitics had centered upon its conflict with Iran, during which it was supported by the United States of America. After the overthrow of the pro-Western Shah of Iran in 1979 by the fundamentalist Islamic regime of Ayatollah Khomeini, Saddam Hussein's war with his Persian neighbor was seen as a means to prevent Iran expanding its influence in the region. The Iran–Iraq war lasted eight years (1980–8). Estimates of the number of casualties vary, but it is thought that the Iranian dead amounted to between 400,000 and 600,000 and the Iraqis suffered approximately 150,000 casualties (Brogan, 1990, p. 263). The war ended in a stalemate, but the situation was enough for Hussein to claim a "victory" to his people. To maintain his political legitimacy, Hussein emphasized border disputes with Kuwait but used the rhetoric of Arab nationalism.

Iraq and Kuwait had been part of the Ottoman Empire, but Great Britain had established de facto protectorate control over Kuwait before the final Ottoman collapse. At the end of the nineteenth century Britain's policy was to control the Persian Gulf, and though it had no real interest in controlling Kuwait its policy was to exclude other powers. Of special concern to the British was the influence of the Ottoman allies Germany and, to a lesser extent, Russia. What was at issue was the discussion of a railway from Baghdad to the Gulf, which was seen by the British as a means of extending German influence in the region. The key question was where on the Persian Gulf coast would the railway end? To cut a long story short, Britain aimed to determine the location of the railhead by defining Kuwait's borders in such a way as to control the mouth of the Shatt el Arab and make Basra, in Ottoman Iraq, an unattractive choice for the rail terminal (Schofield, 2003).

But the view from Iraq was different, and remained a viable political stance for Iraqis through the twentieth century. In Hussein's interpretation Kuwait was part of the Ottoman province under the authority of Basra. In this logic, Kuwait should be part of Iraq as Kuwait had been part of the Ottoman territory controlled by what is now an Iraqi city. Both the logic and the historical interpretation were false. Ottoman control of Kuwait had not really ever existed, and it was outmaneuvered in its political claims to the territory by the British. However, historical record was not the concern of Saddam Hussein, his targeting of Kuwait rested more upon his interpretation of a contemporary material concern, oil (Schofield, 2003; Slot, 2003).

Kuwait possesses the fourth largest oil reserves in the world, behind Saudi Arabia, Iraq, and Iran (World Oil, 2003). In the rhetoric of Hussein, Kuwait "owed" Iraq because of the service it had performed in fighting a war for all Arabs against Iran. With $40 million of debt from the war, Hussein was looking for political glory by receiving thanks for his efforts from his Arab brothers, but also a fraternal injection of cash to his war-ravaged economy. Neither was forthcoming. And so, Hussein portrayed the Kuwaiti regime as misusing their oil fortune. Rather than sharing oil wealth among the Arab nation, in the form of a regional sharing of wealth, the Kuwaiti elite were, according to Hussein, taking the money out of Kuwait and spending it in an immoral manner. By invading Kuwait, Hussein asserted that he would be claiming Kuwaiti oil reserves for the benefits of all Arabs. In reality, if the invasion had been successful Iraq could have dominated the world's oil production.

The material aspect of Hussein's geopolitical code was domination of the Persian Gulf oil reserves in order to counter the costs of the Iran–Iraq war and provide the economic basis for dominance in the region. These actions were justified, with special focus on the whole Arab world, by framing material concerns within Arab nationalist rhetoric and language of religious identity and conflict in the name of Islamic brotherhood (Figure 4.5).

One ingredient of Hussein's message was Arab unity in which oil wealth was portrayed as a common resource being stolen by the wealthy and immoral emirs (Long,

بريشة الفنان محمود حمد

Figure 4.5 Saddam Hussein: benefactor of the Arab world.

2004, p. 29). Hussein claimed he would facilitate an Arab nation that "will return to its rightful position only through real struggle and holy war to place the wealth of the nation in the service of its noble objectives" (quoted in Long, 2004, p. 29). In a clever turn of phrase, this was an internal Arab conflict, a *thawra* (revolution) against *tharwa* (wealth) (Long, 2004, p. 29).

President George H.W. Bush's efforts to establish a UN approved, but US-led, coalition to oust Iraq from Kuwait provided another rhetorical angle for Hussein. As US troops were stationed in Saudi Arabia, Hussein began to portray the conflict as a religious one; Islam was under threat from Zionism and colonialism. This clever move by Hussein added the plight of the Palestinians into his message, painting a picture of his geopolitical code as a defense of the most downtrodden of all Arabs, rather than a grab of oil reserves. Also, it turned the coming conflict with US forces into a defense of Islam against a new "crusade" with, of course, Hussein and the Iraqi people at the vanguard. Hussein was portraying Iraqi geopolitics in terms of pan-Arab social justice, the long-awaited stand against Zionism, and a defense of the Islamic religion.

On December 10, the Iraqi Ba'thist party released the following communiqué, which leaned heavily toward the secular language of Arab nationalism:

> Masses of our militant Arab masses, the Arab Socialist Ba'th Party, which considers this pan-Arab confrontation the Arab nation's battle, calls on you to entrench military cohesion ... [and] to assert the reality of the dialectical relationship between Iraq's steadfastness and the inevitability of its victory in the crucial battle on the one hand and the intifadah of the occupied territory and the liberation of Palestine ... on the other.
>
> (Quoted in Long, 2004, p. 92)

On August 24, Baghdad Radio gave a special broadcast entitled "Muslim Unity Needed" in which the following geopolitical claims were made:

> From the religious point of view a Muslim cannot ask for help from a non-Muslim under any circumstances but the infidel Saudi ruler asked for the protection of Israel and the US which are both not only non-Muslim but also hold a grudge against Islam. It is the same old imperialist ways of intervention in an occupation of others' territories. The world cannot forget the American crimes in Hiroshima and Nagasaki nor can the world forget the lesson taught to the Americans in Saigon, Vietnam, and Korea where the American's poison gas and napalm proved to be of no effect.
>
> (Quoted in Long, 2004, p. 106)

Iraq's forces were defeated overwhelmingly by the number and technical superiority of US-led forces. While a ticker tape parade welcomed home the commander General Schwarzkopf, coalition forces were slow in reacting to Hussein's brutal efforts to cling on to his authority. Uprisings in Shia areas in the south and Kurdish regions in the north were suppressed brutally. In the wake of disturbing pictures of Kurdish refugees, the coalition eventually acted through the imposition of "no-fly zones" that constrained the geographic reach of Hussein's regime.

Part of Hussein's rhetoric in the first Gulf War focused upon his ability to fire missiles, with chemical warheads, into Israel. A few missiles were fired and fell on Israel, but they contained no poison, provided no real threat, and mainly served an ideological goal in illustrating Hussein's role as liberator of the Palestinians. Memory of those weapons loomed large in interpretation of Hussein's denial of UN weapons inspectors. In what is now known to be partial and faulty intelligence, the US government, backed by Great Britain, claimed that Hussein was building a chemical and, perhaps, biological arsenal of weapons that would threaten Israel and US troops in the region. War came again, and this time Hussein was overthrown.

Case study 4.2: United States representations of the 2003 invasion of Iraq

The reasons given for the 2003 US invasion of Iraq rested upon claims of territorial security. The Bush administration alleged that Saddam Hussein had links with al-Qaeda and that he had also developed "weapons of mass destruction" to such an extent that they could be deployed in 45 minutes, as Prime Minister Blair told the House of Commons. These allegations and the subsequent retractions are a matter of political history. For our purposes, the invasion of Iraq and the War on Terrorism in general illustrates points that are germane to our discussion of the geopolitics of world leadership and the representation of geopolitical codes.

The key point is that world leaders must represent their military actions in a manner that is different from other countries. First, they must justify their global role. In other words, the need and ability of world leaders to exert influence in other countries becomes a generally taken for granted feature of world politics. In the case of the invasion of Iraq, the question was whether the US had the specific *grounds* to invade, but the *right* for the US to intervene in another country was not questioned. If the evidence was there to invade, then invading seemed a proper thing to do. Second, the basis for invasion must be linked to the moral message of world leadership.

In numerous speeches and press releases after the terrorist attacks of September 11, 2001, the right, or assumed duty, of the United States to act globally was reasserted. The world was seen as harboring threats that required a global presence. However, the justification and methods of such a global presence were not portrayed in terms of power politics, for that would emphasize material interests. Instead, values were emphasized, or in other words, the global actions of the world leader were portrayed as benevolent actions that would benefit all. The following quotes are from Flint, 2004 (see also Flint and Falah, 2004):

> We will rid the world of the evil-doers. We will call together freedom loving people to fight terrorism.
> (www.whitehouse.gov/news/releases/2001/09/20010916–2.html)

> Terrorists try to operate in the shadows. They try to hide. But we're going to shine the light of justice on them.
> (www.whitehouse.gov/news/releases/2001/10/20011010–3.html)

> Eventually, no corner of the world will be dark enough to hide in.
> (www.whitehouse.gov/news/releases/2001/10/20011010–3.html)

> The US is the defensor [*sic*] of liberty all over the world, and that's what this
> attack was about.
>
> > (www.whitehouse.gov/news/releases/2001/10/20011015–3.html)

> There is no corner of the Earth distant or dark enough to protect them. However
> long it takes, their hour of justice will come.
>
> > (www.whitehouse.gov/news/releases/2001/11/20011110–3.html)

The theme of "justice" was used to make the argument that the War on Terrorism
was part of the project of world leadership. Attorney General John Ashcroft (more akin
to the British Cabinet position of home secretary) was quick to equate the United States
with "civilization" and "justice," and to make these universal ideas part of the global
mission of a particular country, the US.

> A calculated, malignant, and devastating evil has arisen in the world.
> Civilization cannot ignore the wrongs that have been done. America will not
> tolerate their being repeated. Justice has a new mission—a new calling against
> an old evil.
>
> > (John Ashcroft, Attorney General Ashcroft and Deputy Attorney General
> > Thompson Announce, Reorganization and Mobilization of the Nation's
> > Justice and Law Enforcement Resource, November 8, 2001
> > www.justice.gov/ag/speeches/2001/agcrisisremarks11_08.htm)

If the justification of the War on Terrorism lay in providing global justice and
order, then there must be related practical actions. Secretary of State (akin to the
British position of foreign secretary) Colin Powell, began to define what the actual manifestation of the innovations of world leadership would look like. There are strong
similarities between this statement and NSC-68, which was introduced in the previous
chapter. A threat has been identified, in this case a very vague "terrorism," that will be
addressed by seemingly universal notions of "justice" and "civilization" but are in reality the visions of one particular country, and requires a global commitment to building
particular institutions and ways of practicing politics, teaching, economics within other
countries:

> But the war on terrorism starts within each of our respective sovereign borders.
> It will be fought with increased support for democracy programs, judicial
> reform, conflict resolution, poverty alleviation, economic reform and health and
> education programs. All of these together deny the reason for terrorists to exist
> or to find safe havens within those borders.
>
> > (Remarks to UN Security Council, Secretary Colin L. Powell.
> > New York. November 12, 2001, www.state.gov/)

Though the justification for this geopolitics of world leadership is couched in the
establishment of particular "innovations," it is built upon a foundation of military might
and power. Furthermore, this power is seen as ultimately serving national interests,

which implicates the rest of the world. Secretary of Defense Donald Rumsfeld couched the realities of the geopolitics of world leadership in his own inimitable style:

> We talked early on, the president did, about the opportunity to rearrange things in the world in a way that would be beneficial to our country and to peace and to stability and to free systems, and how as we're doing this do we do it in a way that because it's such a fundamental shift in how people think about the world, how do we do it in a way that benefits the world after this event is over.
>
> (Secretary of Defense Donald H. Rumsfeld, Wednesday,
> January 9, 2002. Interview with the *Washington Post*,
> www.defenselink.mil/news/Feb2002/t02052002_t0109wp.html)

Perhaps one could reflect on the images we discussed a little earlier broadcast by Al-Jazeera, or the description of Baghdad by the blogger "Riverbend." Apparently, what they saw was the outcome of "the opportunity to rearrange things." More generally, and with regard to the topic of representing the geopolitics of world leadership, the last quote encapsulates the themes from the other members of President George Bush's cabinet at the time: there is no need to justify the global role of the United States, it is assumed to be "natural"; geopolitical actions under the guise of world leadership do provide material benefits for the United States; such self-interest is equated with benefits for the whole world; such benefits are given and defined through values; the implementation of world leadership requires the construction of institutions and programs within other countries; the fact that the mission of world leadership also requires a military presence and is being actively resisted does not result in self-examination but is seen as an opportunity for further global interventions.

Summary and segue

In the conclusion to this chapter it is important to emphasize that we have discussed representations rather than "facts." If the calculations for war can be traced to material interests, such as access to oil, governments must usually emphasize values or ideas in justifying their foreign policy, especially when it involves invading a country rather than defending one's own. Two important audiences must be addressed to justify a country's geopolitical codes: the domestic and the international audience. The world leader has a particular burden when it comes to representing its geopolitical code, it must convince the whole world that it is acting for the benefit of all rather than for its own interests and gain.

An understanding of geopolitical codes and their representation need not be interpreted within Modelski's model of world leadership. Geopolitical codes may simply be seen as a way of clarifying the decision-making process all countries must make. However, placing the construction of geopolitical codes within Modelski's model allows us to contemplate two things. First, there is a structural explanation for why the agency of states varies over time, and hence why their codes and means of representing them will change. As a country progresses through the phases of the cycle of world leadership, from war to deconcentration, the opportunities and constraints upon a country will alter in a way that requires a dynamic geopolitical code. Second, the inequities of power that are inherent in Modelski's model suggest that the constraints and opportunities for defining the content of geopolitical codes will vary across a hierarchy of power. As the last section of this chapter illustrated, the world leader has particular tools and needs in representing its global mission. Another group of countries, such as Nazi Germany, will define their geopolitical codes in such a way as to place a challenge to the existing world leader. Also, in a broad category that includes varied and specific responses, most countries are too weak to challenge or act as world leader. Their codes reflect either the choice to be compliant with the world leader's project or to resist it.

So far we have discussed the agency of countries within the structure of world leadership. Though the existence of other geopolitical agents has been acknowledged, countries remain the dominant agents. In the next chapter we stop using the word "countries" and develop a more complex understanding of these geopolitical agents.

Having read this chapter you will be able to:

- understand that popular culture is an integral part of geopolitics;
- critically evaluate government statements justifying foreign policy;
- relate the use of different justifications used by countries to the different geopolitical situations;
- think critically about the way other countries are represented in popular culture;
- think critically about the way the foreign policy of one's own country is portrayed in popular culture and government statements.

Further reading

Hedges, C. (2003) *War is a Force That Gives Us Meaning*, New York: Anchor Books.
  An exploration, sometimes disturbing, into the way that our individual and collective identities are inseparable from the practice of warfare.

Said, E. (1979) *Orientalism*, New York: Vintage Books.
  The seminal text on the representation of "others" in the popular media.

The following readings provide examples of how the representational strategies of geopolitics have been studied.

See the special issue of the journal *Geopolitics* 10 (Summer 2005) on geopolitics and the cinema edited by Marcus Power and Andrew Crampton.

Dodds, K. (2003) "Licensed to Stereotype: Geopolitics, James Bond, and the Spectre of Balkanism," *Geopolitics* 8, pp. 125–56.

Flint, C. (2004) "The 'War on Terrorism' and the 'Hegemonic Dilemma': Extraterritoriality, Reterritorialization, and the Implications for Globalization" in O'Loughlin, J., Staeheli,

L., and Greenberg, E. (eds) *Globalization and Its Outcomes*, New York: The Guilford Press, pp. 361–85.

Flint, C. and Falah, G.-W. (2004) "How the United States Justified its War on Terrorism: Prime Morality and the Construction of a 'Just War'," *Third World Quarterly* 25, pp. 1379–99.

Gregory, D. (2004) *The Colonial Present*, Oxford: Blackwell Publishing.

McAlister, M. (2001) *Epic Encounters: Culture, Media, and US Interests in the Middle East, 1945–2000*, Berkeley, CA: University of California Press.

Sharp, J. (2000a) *Condensing the Cold War:* Reader's Digest *and American Identity*, Minneapolis, MN: University of Minnesota Press.

EMBEDDING GEOPOLITICS WITHIN NATIONAL IDENTITY

In this chapter we will:

- define the terms nation, state, and nation-state;
- discuss nationalism and its different manifestations;
- discuss the way gender roles are implicated in nationalism;
- make connections between gender roles, nationalism, and geopolitical codes;
- provide a classification of different nationalisms and their impact upon geopolitical codes;
- provide a brief case study of an ongoing nationalist conflict: Chechnya.

(Misused) terminology

It is important to get our terminology correct, before we proceed. Up to now, I have mainly used the term "country" when referring to the United States, Great Britain, Iraq, etc. The more precise term is state. This can be confusing, especially in the US, where the term state is used to refer to the fifty separate entities that comprise the country. While discussing geopolitics, however, it is more precise to refer to countries as states. Hence, the United States is actually a state, as is Great Britain, Kuwait, France, Nigeria, etc.

States are defined by their possession of sovereignty over a territory and its people. States are the primary political units of the international system. A state is the expression of government control over a piece of territory and its people. The geographic scope of the governmental control exists in a series of nested scales. For example, the London Borough of Hackney is a scale of government, nested within the Greater London Council, the United Kingdom, and the European Union. In another example, the Borough of Queens is a scale of government within New York City, New York State, and the United States of America. We would refer to Hackney and Queens as the local state or local government. In this book, state means the country: for more discussion on the political geography of states see Cox (2002) and Painter (1995).

The terminology may be confusing because it is so widely misused. Instead of using the term state, the term nation or national is usually substituted. Hence, we stand for the national anthem, rather than the state anthem, in the World Cup and Olympic Games we say that national teams compete, rather than state teams. However, the term nation has a very specific meaning that, if we focus on the definition, should not be used in this way. A nation is a group of people who believe that they consist of a single "people" based upon historical and cultural criteria, such as a shared language. In some contexts, membership of a nation will be granted only if inheritance, or blood ties, to members of a particular group can be established, but most nations do not require such blood ties. An introductory discussion of nations and nationalism can be found through the work of A.D. Smith (1993 and 1995).

The geopolitics of nationalism 1: constructing a national identity

Dulce et Decorum Est pro Patria Mori? (See Box 4.1.) Can people be motivated to kill and die for a government bureaucracy? It is hardly a sense of attachment to the Ministry of Defense or the State Department that inspires people to fight. Instead, the ideology of nationalism has equated national well-being with control of a state, the state and nation become synonymous, and the sense of identity is focused upon the nation rather than the state. Nationalism is the belief that every nation has a right to a state, and, therefore, control of a piece of territory. The ideology of nationalism claims that a nation is not fulfilled, the geopolitical situation is perceived to be unjust, if a nation does not have its own state. The geopolitics of nationalism has resulted in millions of deaths, as people fought to establish a state for their nation, and defend their states, in the name of national defense, against threats, real and perceived (Figure 5.1).

The state is equated with the nation through another term, the nation-state: the notion that each state contains one nation. Hence, the Australian nation-state, for example, refers to an Australian nation contained within the Australian state. Such is the ideology. The reality is much different, and the potential for conflict is large. Nearly all states have a diverse population of cultural groups: some of which may define themselves as separate nations (Gurr, 2000). In some situations, a national identity may take precedence over an ethnic identity (Arab-Americans or Italian-Americans, for example). In other cases, a group may demand a degree of autonomy, especially in terms of cultural practices such as the use of language in schools. When a cultural group defines itself as a nation, often there are demands for a separate state for that nation, the politics of nationalism. We will look at the politics of creating nation-states in two ways: top-down and bottom-up.

Top-down nationalism refers to the role of the state in creating a sense of a singular, unified national identity (Mosse, 1975). The United States is, perhaps, the best example of this process. The history of the United States defines its national identity as an immigrant nation: a collection of individuals from national groups across the globe. The practice of the state has been to ensure a centripetal political force: that such a collection of nations does not create conflict, but is "a more perfect union," an American

The King commands me to assure you
of the true sympathy of His Majesty and
The Queen in your sorrow.

He whose loss you mourn died in the
noblest of causes.   His Country will be
ever grateful to him for the sacrifice he has
made for Freedom and Justice.

Milner

Secretary of State for War.

Figure 5.1 World War I telegram to next of kin.

nation. Education is the vehicle for this continual process of creating a nation-state. Children pledge allegiance to "the flag" at the beginning of each school day. The American nation is celebrated in song, dance, and study—the mythology of the nation, the sense of unity, and the child's place within it is created in a "banal" or everyday manner (Billig, 1995). The celebration of the American nation (and also the Australian and Canadian national histories) illustrates that there are positive interpretations of nationalism as a collective identity that transcends ethnic differences.

Ironically, the dominant mythology of the United States as an immigrant land of opportunity rests upon a history in which different cultural groups have suffered at the hands of the state: racist immigration policies targeting Chinese, such as the Chinese Exclusion Act of 1882, the near genocidal Indian Wars of the 1800s, as well as the enslavement of black Africans and the African-Americans' struggle for civil rights that continues today, and the contemporary harassment of Arab-Americans at airports and other security points in the name of the War on Terrorism. However, the power of the United States national identity is that despite the discrimination that successive waves of immigrants have experienced, and still do, the desire to be part of the American nation is still strong, and the degree of assimilation is high, compared to other countries.

The top-down nationalism of the United States illustrates the way the state apparatus has been brought to bear to create a nation. It is a form of nationalism; it promotes the ideology that the state is the natural and obvious political geographic expression of a singular nation. Funnily enough, it is not the type of politics we usually think about when we label a politician a "nationalist": such terminology is usually a form of epithet referred to "monsters" such as Slobodan Milosevic or Adolf Hitler, for example.

Activity

How and where did you learn your national history?

- Think of the settings (home, school, etc.) where you were exposed to this history and the form the history took (books, films, lessons, etc.).

- Write down two or three of the key ingredients of the history and what "moral" or story they may say about the particular national character.

- Write down two or three key historic events that are usually ignored or played-down in the national history of your country.

- How do these lesser discussed events contradict the portrayal of the national character you identified from the dominant narrative?

The geopolitics of nationalism 2: the process of "ethnic cleansing"

Let us turn to the politics of violent nationalism, or bottom-up nationalism, now: the type of nationalism that makes the headlines (Dahlman, 2005). Nationalism, in this sense, is the goal to create a "pure" nation-state, in which one and only one culture or national group exists. This geopolitical perspective views a nation-state as somehow tainted, weak, a geopolitical anomaly, if it contains multiple nations or ethnicities. Instead of the politics of assimilation, the geopolitics here is of expulsion, and eradication. Bottom-up nationalism is what has become known, almost nonchalantly, as ethnic cleansing. Though it is the bloody actions of ethnic cleansing (the killing and rape) that is the "sharp-end" of this form of nationalism, the way the term has become readily,

Figure 5.2 Prelude to ethnic cleansing.

and quite uncritically, adopted by mainstream media as a handy phrase to "make sense" of an event also shows that we are implicated too. As viewers, the pervasive ideology of nationalism makes the goals of "ethnic cleansing" understandable: it is, simply, the most extreme form of the politics of exclusion that underlies discussion of immigration and refugee policies in "civilized" debates in the British parliament, for example. The politics of otherness related to particular territories is the underlying geopolitics.

The process of "ethnic cleansing" can be illustrated schematically. In the first diagram (Figure 5.2), two neighboring states are both multi-national: Triangle State is populated mostly by people △ with a scattering of people ○ near the border, and Circle State displays the opposite pattern. Also, there are both △ and ○ people living outside the borders of these two states. The existence of people of different nations does not

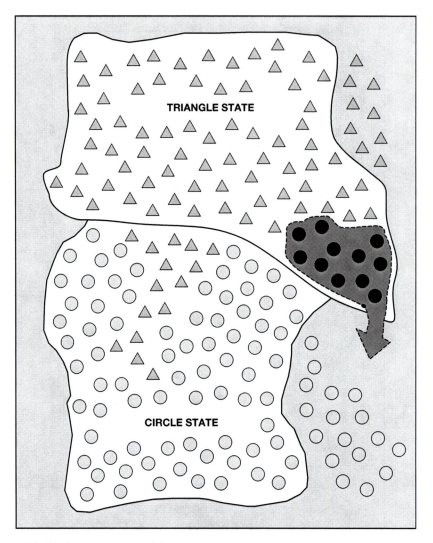

Figure 5.3 Ethnic cleansing: expulsion.

determine conflict; most states exhibit this mixture of nations without violence. But, in some cases, politicians gain prominence on the back of calls to alter the multi-national make-up of the state, often by blaming economic and social woes upon the presence of a minority nation.

The drive to create a "pure" nation-state is illustrated in Figure 5.3. In what has become known as "ethnic cleansing" the minority ○ nation is expelled from the land-scape of the Triangle State. Expulsion usually consists of violence against people and their property that forces them to flee for their safety, leaving their property and posses-sions behind. Expulsion usually takes place in conjunction with eradication (Figure 5.4); the slaughter, usually of young men, and the rape of women to prevent the reproduc-tion of future generations. Rape is also a powerful weapon (see Chapter 8). Women are

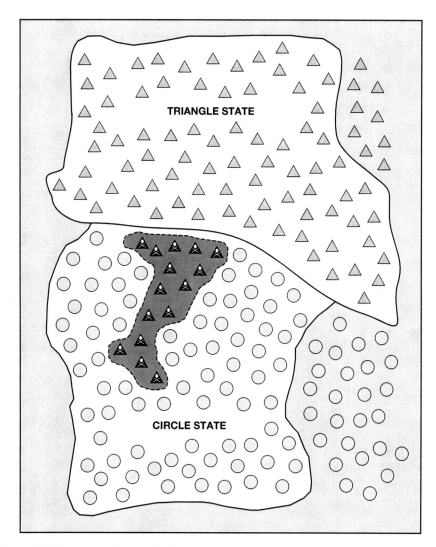

Figure 5.4 Ethnic cleansing: eradication.

"defiled" in order to "pollute" the purity of the nation. In our schematic example, Circle State has retaliated to the expulsion of people ○ in Triangle State by killing people △ within the borders of the Circle State. The goal of ○ and △ nationalists is the creation of a state containing one, and only one, national group.

Sometimes, the geopolitics of nationalism will stop here. In other cases, there may be a further violent step (Figure 5.5). After the bout of ethnic cleansing, Circle State may be pure, but for some the process is incomplete: not all of the members of nation ○ reside within the borders of Circle State. In a fundamentalist interpretation of nationalism, the members of nation ○ in Triangle State are unfulfilled, denied their right to participation in the ○ nation-state. The result is the mobilization of force to change the borders of Circle State so that all members of nation ○ now reside within it. At the

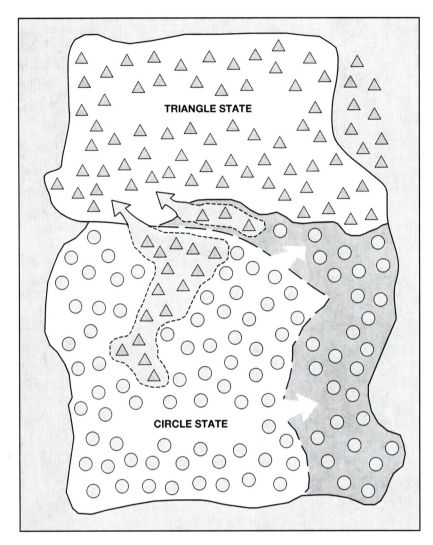

Figure 5.5 Ethnic cleansing: expansion.

same time, any △ people in Circle State who survived the killing flee or are expelled. The "purity" and "wholeness" of nation ○ has been achieved, a lot of blood has been spilt.

Case study 5.1: Chechnya

Conflict in the breakaway Russian province of Chechnya has been an ongoing and brutal process of nationalist geopolitics. The struggle is about, on the one hand, the territorial integrity of Russia and, on the other hand, the desire by Chechens for their independent country (Figure 5.6). In other words, there is a fight over the establishment of Chechnya as an independent nation-state. The population of Chechnya is approximately one million and composed primarily of Chechens and Ingushes as the dominant ethnic groups in the region. Most Chechens are Sunni Muslims. Chechen and Russian civilians alike have suffered in the conflict. Approximately 200,000 to 250,000 Chechens have been forced to leave their homes, most of them during the Russian invasion of 1999, while Russian civilians have had to live in fear of suicide bombings and various other forms of terrorist attacks throughout Russia. For more information on the conflict refer to the readings listed at the end of this chapter and the specific sources cited. The purposes of this case study are:

- to provide historical background to help you interpret a recurrent geopolitical conflict;
- to illustrate how nationalist conflicts are viewed differently by different social groups;
- to offer a concrete example of the concepts discussed in the chapter.

History of the conflict

1893   While industrialization was sweeping over Russia, oil was discovered in Chechnya (which was at this time a part of Russia)—the area became increasingly important to Russia.

1890s   Russia built the Vladikaukaz railroad line through Chechnya—Chechnya was a key route to southern regions of Russia.

1914   By this time, Chechen oil comprised 14 percent of Russia's oil production.

1917   Beginning of the Bolshevik Revolution—Tsar Nicholas II and then the Provisional Government were both ousted by the Bolshevik Party.

1918   The Mountain Republic was established, only to be taken back by Lenin in 1921.

1922   The Mountain Republic was officially dissolved into the Chechen Autonomous oblast on November 30.

1923   Lenin's Congress officially adopts the policy of *korenizatsiya* (indigenization), encouraging different nations to use their languages and celebrate their cultures: instilling a sense of ethnic and national awareness in minority groups.

1934   The Chechen Autonomous oblast merged with the Ingush Autonomous oblast.

1936   The combined oblast rises to the status of Autonomous Soviet Socialist Republic (ASSR).

1944   Stalin begins deporting Chechen and Ingush people to Siberia and Central Asia, accusing them of conspiring with Nazis.

- "On February 23, 1944 over 500,000 Chechens and Ingush were transported to northern Kazakhstan for an exile that lasted 13 years" (German, 2003, p. 4). Chechen language publications were banned and the term "Chechen" [plus descriptive terms of other nationalities that had been deported] was erased from Soviet textbooks and encyclopedias.

1957   January—Nikita Khrushchev re-establishes the Chechen-Ingush ASSR.

- "The return of the deportees did aggravate tensions in this region, which could not support an influx of people who were without homes or employment" (German, 2003, p. 19).

1985   Gorbachev rises to power.

- "The advent of glasnost and perestroika [the political and social changes that signaled the collapse of the Communist Party's control of the Soviet Union], and consequent relaxation of previous restrictions, heralded the appearance of popular fronts demanding greater autonomy for the manifold ethnic groups" (German, 2003, p. 14).

1991   The collapse of the Soviet Union.

- November 1—Dzhokhar Dudayev, after winning a presidential poll, proclaims Chechnya independent of Russia.
- Russian President Yeltsin declares Martial Law in Chechnya and Ingushia, sending 1,000 Internal Affairs Ministry troops—but they leave without ever disembarking from their aircrafts as crowds block the airport.
- Chechnya begins to develop its army. It is to consist of 11,000–12,000 troops with approximately 40 tanks and 50 units of various armored equipment.

1990   November—Former Communist leaders "convoked a Chechen National Congress . . . and invited the recently promoted General Dudaev—who had never lived in Chechnya—to head the nationalist movement" (Evangelista, 2002, p. 16).

1992   Chechnya adopts a Constitution.

1994   Russian troops invade Chechnya to end the independence movement.

- During the 20-month war that follows, approximately 100,000 people, many of them civilians, are killed.

1995   Chechen rebels seize hundreds of hostages at a hospital in Budennovsk, southern Russia. Over 100 are killed in the initial raid and the subsequent unsuccessful Russian commando operation.

1996   April—President of Chechnya, Dudayev, is killed by a Russian missile attack. Succeeded by Zemlikhan Yandarbiyev.

May—Yeltsin signs short-lived peace agreement with Yandarbiyev.

August—Chechen rebels attack Grozny (rebel chief of staff Aslan Maskhadov and Yeltsin's security chief Alexander Lebed sign the Khasavyurt Accords—ceasefire).

November—agreement of Russian troops' withdrawals.

1997   January—Aslan Maskhadov wins Chechen presidential elections. His presidency is recognized by Russia.

1999   January—Maskhadov announces his three-year plan to phase in Islamic Shariah Law.

March—Moscow's top envoy to Chechnya is declared missing (his body is found in Chechnya one year later).

July/August—Chechen insurgents begin crossing the border into Dagestan to assist the overthrow of the Russian government and the establishment of a separate state. Maskhadov tried to maintain friendly relations with Russia and appealed to Chechens to leave Dagestan.

September—a series of bombings targeting Russian civilians, attributed to Chechen rebels, provoked the Russian government to take action.

- Approximately 300 Russian civilians killed.

October—Russian government launches assault into Chechnya and recaptures the breakaway regions of Dagestan.

- It is estimated that approximately 200,000 refugees flee Chechnya for neighboring Russian republics.

2000   Russian troops capture Grozny. Russia declares rule of Chechnya. War continues in mountainous areas.

June—Akhmat Kadyrov is named the head of administration in Chechnya by the Russian government.

2001   January—rumors of human rights atrocities begin to circulate in the Chechen village of Dachny.

February—the first body found in a mass grave in Dachny.

- After later inspection, it was reported that 48 bodies were found, 34 of which were never to be identified.

2002   October—Chechens seize an 800-person theater in Moscow.

- One man shot and killed by the hostage-takers.

- 118 civilians (along with 50 hostage-takers) die as a result of gas used by the Russian government to flush out the terrorists.

2003   March—Chechens vote in referendum that makes Chechnya a separatist republic within Russia. Multiple suicide bombings occur following the referendum.

Akhmat Kadyrov becomes the official president (many groups question the legitimacy of the elections).

2004   February—former Chechen president, Zelimkhan Yandarbiyev, is murdered in a car bombing in Qatar.

- Two Russian intelligence officers are sentenced to life in prison after admitting that the attack was ordered by the Russian government.

May 9—Akmat Kadyrov and five others are killed in a bombing in a stadium in Grozny. Some 56 people are wounded in the attack.

- "The blast comes a few weeks after Putin, in his annual state of the nation address, proclaimed the 'military phase of the conflict may be considered closed' in Chechnya" (CNN, 2004).
- Warlord, Shamil Basayev, declares responsibility for the attack.

The conflict spreads into neighboring Ingushetia—known to be a peaceful safe-haven.

- "A year ago it was an idyllic farming region but now many locals refer to it as 'a second Chechnya'" (Walsh, 2004).
- Cassette tapes surfaces with a Russian soldier's confession to the kidnapping of a senior aide to the top prosecutor of Ingushetia—this aide was pursuing prosecutions against Russian soldiers who were accused of abuses.

Experiencing the conflict

Anna Politkovskaya is a leading investigative reporter for the Russian newspaper, *Novaya gezeta*. She reported on the conflict, including the plight of refugees, bombed by Russian forces as they fled:

> They fly so low that you can see the gunners' hands and faces. . . . They seem to be laughing at us crawling comically down below—heavy old women, young girls, and children.
>
> (Politkovskaya, 2003, p. 33)

For those who remained in Chechnya, they were targets of both the Russian military and Chechen groups who targeted civilians who worked for or cooperated with the Russian-friendly government. Politkovskaya describes the "pits," where the Chechens are kept to await questioning—in the middle of winter without any other shelter:

> It's about ten by ten. . . . Despite the frost, there is a distinct odor. That's how they do things here—the Chechens have to go to the bathroom right in the pit.
>
> (Politkovskaya, 2003, p. 50)

> An old woman, already a grandmother, accused of harboring militants spent twelve days in the pits. She was taken out, tortured by electric shock, never questioned, and handed over to her family and neighbors after they gathered enough money for ransom.
>
> (Politkovskaya, 2003, p. 48)

Residents of Moscow were hostile to Chechen and Ingushetia refugees, especially after the hostage situation at the Moscow theater in 2002. Human Rights Watch (2004) cites various abuses of ethnic Chechens in Moscow that include: arbitrary identity checks and detention, the planting of drugs or weapons, obstruction of registration in Moscow, harassment of unregistered Chechens in schools, and pressure on landlords to evict

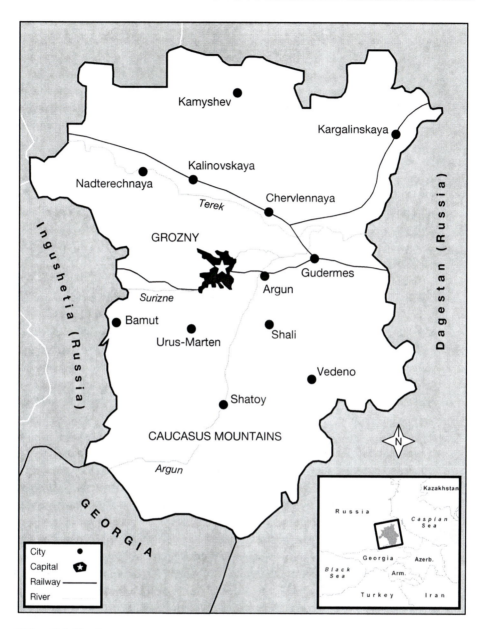

Figure 5.6 Chechnya.

Chechen tenants. Human Rights Watch conducted interviews with victims of various forms of discrimination. One woman told them that she tried to register in Moscow but the registration official told her: "Your wanting to register in Moscow is the same as going to the White House and asking to live there" (Human Rights Watch, 2004).

Even though there is popular hostility to the refugees, Russian citizens were unsupportive of the war. In response, Yeltsin's administration tried to justify the invasion of Chechnya by publicizing Dudaev's most sensational statements in order to instill fear in the Russian people. These included "remarks reportedly made to a Turkish journalist during the war that . . . he would personally fly a bomber to Moscow to retaliate" and that he would kill Russian prisoners of war who tried to assist Dudaev's opponents (Evangelista, 2002, p. 35). However, the development of terrorist attacks has resulted in a greater support for the war among Russian civilians. For example, the bombing of an apartment building that housed only civilians, and other terrorist attacks, had "a traumatic and galvanizing effect on Russian public opinion, comparable to what happened in the United States after the attacks of September 11, 2001. They provoked widespread and uncritical support for expanding the war against Chechnya" (Evangelista, 2002, p. 67). Other actions that galvanized public support were the December 1996 kidnapping of 22 Russian Interior Ministry troops in Chechnya, and, two days later the kidnapping of a government delegation, on its way to Chechnya for talks with Chechen officials. The same night six medical personnel from the Red Cross were shot and killed south of Grozny. The following day six Russian civilians, living in Grozny, were killed (Evangelista, 2002, pp. 46–7).

Russian policy toward Chechnya has been contested, though on the whole the attitude of the political parties is toward a violent end of the conflict to maintain Chechnya as part of Russia. United Russia, a pro-Putin party, hated by Chechen separatists, wishes that Chechnya remain a part of Russia. The Communist Party (of the Soviet Union/of the Russian Federation) also wants Chechnya to remain part of Russia. Yabloko, a more liberal democratic party, was very critical of the war but has toned down its criticisms over time (LaFraniere, 2003). The Union of Right Forces, also liberal, has recently been more critical of Moscow's approach to the Chechen conflict (LaFraniere, 2003).The Liberal Democratic Party, an extreme right-wing party, issued violent ultimatums to the Chechens, threatening their annihilation by military force (Gusev, 1996).

What of the sorry situation of the Russian soldiers? In the Russian military, approximately 3,000 non-combat deaths occur each year: the result of suicide and beatings incurred at the hands of senior ranking officials. Malnourishment is also common (Associated Press, 2003). Given these awful conditions, parents and draft-age youths do everything that they can in order to avoid the draft. "Thousands of young men avoid the twice-yearly call up by obtaining educational deferments, medical and family exemptions, paying bribes, or simply dodging military authorities" (Associated Press, 2003). The Committee of Soldiers' Mothers is a strong force in ending the harsh conditions for soldiers. They vehemently voice their opinion against the status quo and lack of sympathy that exists in the Russian government. The committee does a lot to help parents gain information that has not been readily available in the past. It is involved in "supporting efforts of parents to travel to Chechnya and rescue their sons from the army or at least find out how they died and recover their bodies" (Evangelista, 2002, p. 42).

Popular sentiment in Russia toward the conflict is a combination of animosity toward the Chechens, especially the refugees, but also discomfort over the conduct of the war, and the situation of the draftees. However, this did not stop Vladimir Putin using the US's War on Terrorism as a vehicle to strengthen his commitment to the conflict and justify its continuation. Putin made links between Chechen terrorists and al-Qaeda, saying that some of the fighters were trained in the same Afghan camps. Prior to the beginning of the War on Terror, the countries of the EU were very critical of Russia's handling of the Chechen issue. On September 26, 2001, on a visit to Germany, Putin was promised by Schroeder that criticism would tone down and that there would be a "differentiated evaluation" of the matter (Evangelista, 2002, p. 180). Russia used this opportunity to convince the rest of the world of its shared values. For example, at a meeting of the world economic leaders in February 2002, Russia's Prime Minister Mikhail Kasianov said "Russia understands better than any other nation what has happened in America" (Cullison, 2002).

There are also different views of the conflict from the Chechen perspective. Though a sense of national identity is shared, some, such as, the Chechen leader, Maskhadov, fought for a moderate form of independence—with a peaceful break from Russia, followed by cooperation with Moscow and other regional neighbors. On the other hand, many Chechens believed in a harsh break from Russia, and the construction of an Islamic state "encompassing Chechnya and Dagestan and perhaps other Muslim peoples of the North Caucasus" (Evangelista, 2002, p. 50). However, all the nationalist visions of Chechnya rested upon a shared and dominant history of injustice at the hands of the Russian government, dating from Stalin's regime. Nationalist sentiments have risen from the continued fight with Russia: "The deportations failed to break Chechen resistance and instead contributed to an abiding attachment to the homeland and a smoldering sense of grievances" (Evangelista, 2002, p. 13).

This conflict is likely to continue. Given the role of Russia as ally in the War on Terror, producer of oil and natural gas, and influential power in Central Asia and Trans-Caucus region, Chechnya will be a part of the geopolitical calculations of Russia and other countries. Perhaps, as Politkovskaya (2003, p. 29) notes, the tendency will be to ignore the conflict: "Everywhere they invite me to make a speech about 'the situation in Chechnya,' but there are zero results. Only polite Western applause in response to the words: 'Remember, people are continuing to die in Chechnya every day. Including today.'"

The case study of Chechnya illustrates the tenets of nationalism, where individual and group identity is linked to sovereignty over a particular piece of territory. The nationalist claim is based upon a history that includes injustices, and so promoting the "need" for a protective nation-state. But the geopolitics of nationalism gives rise to conflicts within national politics, as is evident in the broad party support for Russia's policy compared to the disaffection of soldiers and their families. Noting the role of families in nationalism requires us to pay careful attention to the role of gender in nationalist politics.

Gender, nationalism, and geopolitical codes

In *Dulce et Decorum Est pro Patria Mori* Wilfred Owen portrays the harsh world of the battlefield, but it is very clearly a masculine world. In World War I, the roles women played were limited to exhorting the troops to go off to war. Other classic writing on warfare is also a story of men under fire. In the novels *The Naked and the Dead* by Norman Mailer and *The Thin Red Line* by James Jones, women are "back home," to be returned to and remembered, and their potential infidelities a further source of stress. Homosexuality is rarely a topic, though *The Thin Red Line* is a notable exception in its casual recognition that men found comfort with men during warfare.

The masculine nature of geopolitical codes is the goal of our discussion. To get there it is necessary to explore the gendered nature of nationalism, with special reference to the use of nationalism at time of conflict. We began this chapter with an emphasis on "naming" so that we can begin to understand nationalism as a political process of joining a nation and a state. Naming is not just a matter of academic classification; it also refers to the particular label given to an abstract concept, such as "the state," in order to give it popular meaning and salience. Think of other, more colloquial names for the nation. Some that come to mind are homeland, motherland, and even fatherland. Fatherland is noteworthy because this particular gender reference to the nation has a very negative historical connotation: Nazi Germany, or nationalism gone "too far." Instead, we are more comfortable with thinking of the nation as the "motherland," with its references to nurturing, comfort, and sense of belonging. The nation is, as Anderson (1991) noted, an imagined community: meaning that we think of ourselves as part of a national community, but we will never interact with the vast majority of its members in any meaningful way. Instead, the sense of community is "imagined" through national events such as sporting events, elections, the funerals of statesmen and women, natural disasters, etc.

Perhaps a more accurate description of the nation is an "imagined family," and a patriarchal one at that. For notions of the "motherland" imply a particular role for women; they should be active in the procreation and socialization of the nation's future generations, and their domain is the home. The flip side to this gendered role is that men are the seen as the defenders and rulers of the nation—they inhabit the "public spaces" of government, business, and the military. Hence, the dominant narratives of *The Naked and the Dead* and other war novels, the men must fight for and defend their women "back home," but the individual is only complete (and by extension so is the nation) when functioning as a household of a man and a woman: this, the narrative says, is what is to be fought for.

Feminist scholars have shown the gendered division of labor within politics, and foreign policy in particular (see the essays in Staeheli *et al.*, 2004). Despite some positive changes in attitudes toward women, and legal recourse to equality, ideology is harder to change. By ideology I do not just mean the overt sexism of some individuals and political agendas that promote the role of the woman as "homemaker" and or sexual object. Such agendas are by their very brazenness perhaps relatively easy to challenge. More threatening is the insidious or banal practices that promote gender roles that limit women's participation in public space. Nationalism remains an exceptional tool for defining gender roles, and perhaps especially at a time of conflict.

Women and the War on Terrorism

In the United States after the terrorist attacks of September 11, 2001 the words "hero" and "nation" were pervasive and intertwined. Without dismissing the bravery of firefighters, police officers, and others who lost and risked their lives in the wreckage left by the terrorist attacks, it is also striking how gendered this narrative became. Men were the heroes, women the homebound helpless victims. The term "brotherhood" was used repeatedly in references to the New York Fire Department. The bravery of the "brotherhood" of rescue workers on that day was portrayed as a purely masculine pursuit: barely mentioned were the women rescue workers who also served (Dowler, 2005).

Of course, women did play an important role in the nationalist story that was told in the wake of these terrorist attacks; the tragedy of widowhood was exposed to a nationalist light. Again, one must emphasize that an academic analysis of the victims of 9/11 is not intended to diminish or demean individual loss and suffering. However, it is important to see how an individual's loss becomes part of a national tragedy or episode that, in turn, feeds into the redefinition of a geopolitical code, especially its military role. The following excerpt from President Bush's State of the Union Speech of January 29, 2002 is particularly illustrative:

> For many Americans, these four months have brought sorrow, and pain that will never completely go away. Every day a retired firefighter returns to Ground Zero, to feel closer to his two sons who died there. At a memorial in New York, a little boy left his football with a note for his lost father: Dear Daddy, please take this to heaven. I don't want to play football until I can play with you again some day.
>
> Last month, at the grave of her husband, Michael, a CIA officer and Marine who died in Mazur-e-Sharif, Shannon Spann said these words of farewell: "Semper Fi, my love." Shannon is with us tonight. (Applause.)
>
> Shannon, I assure you and all who have lost a loved one that our cause is just, and our country will never forget the debt we owe Michael and all who gave their lives for freedom.
>
> ("President Delivers State of the Union Address"
> (title of webpage www.whitehouse.gov/news/releases/
> 2002/01/20020129–11.html—accessed December 22, 2004)

In this segment of the speech the tragedy suffered by men in the defense of the nation is clearly flagged, in terms of losses suffered "at home" and also "abroad." Shannon Spann's loss was made a public event as television cameras broadcasted her acknowledgment of the applause from the country's elected officials. *Semper Fi* (or *Semper Fidelis*) is the Latin motto of the United States Marine Corps, and is translated as "always faithful": faithfulness that is promised to God, country, family, and the Corps. The use of *Semper Fi* in the State of the Union speech hits upon different aspects of the nationalist story: the women back home are "ever faithful" while the nation's men folk are away fighting (contrary to the constant worries about fidelity of the characters of war novels); such faithfulness is part of the institution of holy matrimony, blessed by a God that is blessing the fighting too; the country demands a loyalty that requires people to put their lives at risk, despite the claims to faithfulness to family. Loyalty to nation and the Corps takes precedence over loyalty to family.

The point to emphasize here is that the practice of geopolitics requires geopolitical agency in many different settings, including the home. The geopolitical actions of states are dependent upon the actions and sacrifice of people like Michael Spann. How can his wife's loss, or the loss of any individual fighting for any country, be justified? In other words, a foundational ideology, applicable in all countries, is necessary to justify the conflict that is undertaken as part of a geopolitical code. Nationalism plays that role by creating a sense of "community" and allegiance that warrants sacrifice.

Gendered nationalism and the masculinity of geopolitical codes

Though each country's nationalism is unique, in the sense of its particular history, nationalism is still consistent in defining particular gender roles, even during what are described as "revolutionary" situations such as Cuba. Women consistently are identified with a subordinate role and their access to public positions limited. In this sense, nationalism can be seen as a structure that contains a message of the "proper" roles for men and women. At times of conflict a state's geopolitical code may reinforce these gender roles as warfare intensifies the different expectations of sacrifice expected of men and women: the "heroic" actions of fighting men and the "stoic" sacrifice of the women who are left behind. There is a constant feedback in this relationship too. As men continue to dominate the public sphere, a masculine perception of the world is maintained, with the inevitable result of aggressive geopolitical codes and wars.

Nationalism requires the construction of difference between the populations of different states. Such difference allows for the construction of "enemies," "threat," and "danger" as part of a states' geopolitical code. These notions are dependent upon a dominant military view in society that commands a particular vision of "a dangerous world" and how to respond. From a feminist perspective the dominant military view rests upon a masculine view of the world: the implication is that individual gender roles, geopolitical codes, and the structure of global geopolitics are connected in practice and ideology.

The concepts of militarism and militarization are related to how geopolitical codes are constructed and what they contain. The core beliefs of militarism are:

(a) that armed force is the ultimate resolver of tensions;
(b) that human nature is prone to conflict;
(c) that having enemies is a natural condition;
(d) that hierarchical relations produce effective action;
(e) that a state without a military is naive, scarcely modern, and barely legitimate;
(f) that in times of crisis those who are feminine need armed protection;
(g) that in times of crisis any man who refuses to engage in armed violent action is jeopardizing his own status as a manly man.

(Enloe, 2004, p. 219)

Militarism is, then, an ideology, a particular view or understanding of society and how it should be organized. It is a different ideology than nationalism, but they are usually found hand-in-hand. Related to militarism is militarization "the multitracked process by which the roots of militarism are driven deep into the soil of society" (Enloe, 2004, pp. 219–20). One of these processes is the way the military is constructed as a masculine institution and war as a masculine enterprise. In this way, the masculine nature of militarization is complemented and enhanced by the gender roles promoted in nationalism, and vice versa (see Box 5.1).

The implications of militarization are individual, national, and global. In terms of the construction of geopolitical codes, militarization is seen as a foreign policy issue because of the dominant influence of the military in forming codes, and equating security with military matters. Most importantly, militarization is especially successful when civilian policy-makers acquiesce to a foreign policy implemented by force (Bacevich, 2005; Enloe, 2004).

The militarization of a geopolitical code rests upon the dominance of men in positions of public office, who are willing to facilitate a foreign policy that rests upon masculine assumptions about individual behavior that are then transferred to geopolitical codes. The essential ideological building block is the masculinity myth: the notion "to be a soldier means possibly to experience 'combat', and 'combat' is the ultimate test of a man's masculinity" (Enloe, 1983, p. 13; Hedges 2002).What it means to be a "man" and effective military operation are mutually reinforcing:

> Men are taught to have a stake in the military's essence—combat; it is suppos-edly a validation of their own male "essence". This is matched by the military's own institutional investment in being represented as society's bastion of male identity. That mutuality of interest between men and the military is a resource that few other institutions enjoy, even in a thoroughly patriarchal society.
>
> (Enloe, 1983, p. 15)

Box 5.1 Foreign policy as a masculine practice

The militarization of any country's foreign policy can be measured by moni-toring the extent to which its policy:

- is influenced by the views of Defense Department decision-makers and/or senior military officers
- flows from civilian officials' own presumptions that the military needs to carry exceptional weight
- assigns the military a leading role in implementing the nation's foreign policy, and
- treats military security and national security as if they were synony-mous

(Enloe, 2004, p. 122)

Consider the foreign policy of your own country. Who is making statements to the media about a particular issue, military officers, the foreign secretary (secre-tary of state) or minister/secretary of defense? Does the Ministry of Defense/Pentagon give regular news conferences?

Combat defines the "man" and also validates the existence of the military. Moreover, combat as a masculine pursuit translates into the importance of the military as a masculine institution that, furthermore, plays a role in the militarization of geopolitical codes. The militarization of geopolitical codes is especially resonant when combat is in progress, defined as "likely," or a recent matter of national history. The foreign policy experience of "combat" defines the identities of individuals (men and women) and, hence, continues the relationship between the construction of individual identities and the form of geopolitical codes.

Combat, constructed as an essentially masculine pursuit, rests upon women in two ways. One is in the practical sense, the exclusion of women from combat duty but their necessary role of "camp followers" (Enloe, 1983): in other words, women play a number of "supporting roles" that are necessary for the military to function. Some of these roles are with the services, such as nursing and clerical work. Other roles are outside the services and even the law. Prostitution is perhaps the most obvious, but also the role of the military or diplomatic wife (Enloe, 1990). Crucial to our connection of militarization, nationalism, and geopolitical codes is the twin needs of women's support services while women's roles are controlled and restricted to prevent "disorder," in the form of women's participation in combat. In the words of Cynthia Enloe:

> This mutuality of interests has the effect of double-locking the door for women. Women—because they are *women*, not because they are nurses or wives or clerical workers—cannot qualify for entrance into the inner sanctum, combat. Furthermore, to *allow* women entrance into the essential core of the military would throw into confusion *all* men's certainty about their male identity and thus about their claim to privilege in the social order.
>
> (Enloe, 1983, p. 15)

Combat defines the man, but man also defines the combat. In other words, by making combat the definition of manliness, and making combat a male preserve, military combat is the defining event of a patriarchal society and its members. "Women may serve the military, but they can never be permitted to *be* the military" (Enloe, 1983, p. 15; emphasis in original). The militarization of geopolitical codes is enhanced as it serves individual goals—making society's boys into men—while also facilitating a dominant role for the military in the definition of a geopolitical code. In turn, those with "combat experience," by definition men, are also privileged in public affairs (Enloe, 2004), a process very evident in US politics.

The militarization of society complements and intensifies the gender roles that are defined by nationalist ideology. Furthermore, in patriarchal societies "combat" has an essential role in the essential identity or purpose of men. Not surprisingly, war is a key ingredient in national myths and interacts with the gendered understanding of public and private roles. Not only are men the "defenders" of the nation, but actual defense is necessary to make a man. With the militarization of society, in addition to the role of "combat" in defining male identity, and the dominance of men, and the military, in public affairs, it is hardly surprising that the necessary construction of *difference* by nationalist ideology is readily "upgraded" into "hatred" and "threat": in other words, war.

In the next section we discuss how different national histories can be related to different geopolitical codes and hence the forms of conflict that particular nationalisms may generate.

A typology of nationalist myths and geopolitical codes

The geopolitical codes of states rest upon the maintenance of their security. On the whole, security is related to the territorial integrity of the state. In other words, geopolitical codes define ways in which the sovereignty of the state must be protected or the state's status and well-being enhanced. Perceived threat of attack upon the citizens of the country requires a geopolitical code attending to boundary defense. Enhancing the status and well-being of a state often requires identifying historic grievances that have denied a country its "rightful" access to a particular set of resources. Consequently, an aggressive geopolitical code may be written that requires the seizure of territory.

Whether the code tends to be more defensive or aggressive, the concepts of sovereignty and territory remain central to the ideology used to justify the geopolitical code. Three types of "historical-geographic understandings" that frame the specific justifications of particular countries have been identified (Murphy, 2005, p. 283):

1   The state is the historic homeland of a distinctive ethnocultural group.
2   The state is a distinctive physical-environmental unit.
3   The state is the modern incarnation of a long-standing political-territorial entity.

These categories are not deterministic, just because two countries possess a historical-geographic ideology emphasizing, say, territorial integrity, does not mean that they are likely to be equally aggressive. The benefit of this classification is that it shows that the justification for geopolitical actions used by a government must be grounded in a national ideology that resonates with the population: it must "make sense."

For example, the continuing conflict between Turkey and Greece is focused upon islands off the west coast of Turkey as well as the divided island of Cyprus. An interpretation of geopolitics emphasizing material pursuits would point to the oil reserves under Turkey's western continental shelf. But what about the justification for the conflict? The Greek government's response to the Turkish invasion of Cyprus in 1974 is replete with allusions to modern Greece's unbroken connection to the ancient Greek Empire. In the words of the Greek Foreign Ministry:

> The name of Cyprus has always been associated with Greek mythology (mostly famously as the birthplace of the goddess Aphrodite) and history. The Greek Achaeans established themselves on Cyprus around 1400 B.C. The island was an integral part of the Homeric world and, indeed, the word "Cyprus" was used by Homer himself. Ever since, Cyprus has gone through the same major historical phases as the rest of the Greek world.
>
> (Quoted in Murphy, 2005, p. 285)

The connection between Greek gods and an estimated 225,000 refugees may appear tenuous to a neutral and objective observer. The point is that going to war could be justified to the Greek public, and gain support, through the usage of this widely held belief in the national history of the country.

More specifically, we can use the historical-geographical understanding of a country's geopolitical situation to suggest broad relationships between national identity and the content of a geopolitical code, though not in a deterministic sense (Murphy, 2005, p. 286):

1   An ethnic distribution that crosses state boundaries is most likely to be a source of interstate territorial conflict where the ethnic group in question is the focus of at least one state's regime of territorial legitimation.
2   A boundary arrangement is likely to be particularly unstable where it violates a well-established conception of a state's physical-environmental unity.
3   States with regimes of territorial legitimation grounded in a preexisting political-territorial formation are likely to have particularly difficult relations with neighboring states that occupy or claim areas that are viewed as core to the prior political-territorial formation.
4   States that are not in a position to ground regimes in any of the foregoing terms are less likely to have territorial conflicts with their neighbors unless there are strong economic or political motives for pressing a territorial claim and state leaders can point to some preexisting political arrangement or history of discovery and first use that arguably justifies the claim.

The first point is the politics of nationalism discussed at the beginning of this chapter. The second point is illustrated by one of the most puzzling contemporary geopolitical tensions: the dispute between NATO allies Spain and Britain regarding the tiny territory of Gibraltar. Once of strategic importance, given its location at the mouth of the Mediterranean Sea, Spain's desire to gain control of Gibraltar is explained by the historical understanding of the physical extent of Spain; a physical geography currently violated by Britain's possession of just two-square miles of territory. The third point was already exemplified in a discussion of the Turkish-Greek conflict over Cyprus. In addition, China's numerous territorial claims in East and Southeast Asia that rest upon the geographic extent of the ancient Chinese Empire are a contemporary example of the third point.

The final scenario of geopolitical conflicts illustrates an important point: many states must legitimate their geopolitical codes without recourse to a national understanding of political geographies of ethnicity, physical extent, or historical claims. The states of sub-Saharan Africa are colonial constructs; they are recent creations with little basis in ethnic homogeneity or physical legacies. Hence, there has been little cause for border conflicts in this region of the world, instead the geopolitics has been a matter of which ethnic group is able to seize control of the state apparatus, and not the geographical extent of the state (Herbst, 2000). In contrast, the imposed borders of Latin America shifted over the course of Spanish colonialism, creating opportunities for disagreement over their "proper" course.

Geopolitical codes are not simply an objective or strategic calculation made by foreign policy elites; it is not a matter of "statecraft" that excludes the majority of the population. Everyone is implicated, to a certain degree, because geopolitical codes cannot be enacted unless the majority of the population is acquiescent, at least tacitly. To ensure that a geopolitical code resonates with its citizens, a country is careful to frame its actions within the established political geographic sentiment of the nation's history. We have identified some broad categories with which we can interpret the major theme of the national tradition being evoked.

We conclude this chapter by exploring the relationship between national identity and state geopolitics (Dijkink, 1996). National identity frames the geopolitical acts of states within its commonly understood history. Emphasis is placed upon the important role that context plays in determining how and what people believe: "Identifying with a territory simply elicits certain views on the world, albeit in a contingent way, given certain national challenges, historical facts, and ideals" (Dijkink, 1996, p. ix). In other words, "to live within a territory arouses particular but shared visions (narratives) of the meaning of one's place in the world and the global system" (Dijkink, 1996, p. 1). People are socialized within different territorial settings; what they hear, how they make sense of the information they receive, and the possible responses are limited by geographically specific institutions (Agnew, 1987). Referring back to our discussion of place in Chapter 1, it is the uniqueness of a country's geopolitical location or strategic situation, coupled with the way history is interpreted through dominant institutions, which formulates the particular ingredients of national ideology. Even in the age of satellite communication technology, and "globalization," information is distilled and interpreted through local journalistic and government lenses (Dijkink, 1996, p. 3).

Visions of one's country and its position in relation to other countries are formed within particular national myths. These myths form the basis for geopolitical codes and the means to represent and interpret these goals so that they obtain popular support. Dijkink's term is *geopolitical vision*: "any idea concerning the relation between one's own and other places, involving feelings of (in)security or (dis)advantage (and/or) invoking ideas about a collective mission or foreign policy strategy" (Dijkink, 1996, p. 11). National histories are replete with the memories of both the pain of historical suffering and humiliation and the pride of past glories.

It is the tension between how these "maps of pride and pain," as Dijkink calls them, are remembered and used to initiate and justify foreign policy that make geopolitical visions the way national sentiments are translated into geopolitical codes. "[N]ational identity is continuously rewritten on the basis of external events: and foreign politics does not mechanically respond to real threats but to constructed dangers" (Dijkink, 1996, p. 5). Strategic concerns about resources and economics, and ideological referents to national values combine in geopolitical visions, a framing of the world that connects the individual's sense of identity to global geopolitics through the geopolitical code of their country. The content of national myths and the content of geopolitical codes are made within dynamic contexts of conflict. The connections between national myths and geopolitical codes identified by Murphy and Dijkink show that geopolitical conflicts must be understood by connecting the actions of one set of geopolitical agents (those who control the state) with another group of geopolitical agents, the population of those states.

Activity

Identify which of the "historical-geographic understandings," described at the beginning of this section (p. 125), best fit the way the nationalism of your country is portrayed.

- Think of current and past conflicts in which your country has been involved.

- Do the reasons and justifications of these conflicts follow the expectations of the framework? Why?

Summary and segue

Nationalism is an ideology that defines an overarching national identity that transcends ethnic differences. Simply put, the claim is that people within the boundaries of a state hold a common identity. In other words, the emphasis is upon homogeneity. Increasingly, however, emphasis is being placed upon hybridity. Individuals possess multiple collective identities, and, furthermore, these collective groups are themselves the product of mixture. The identity Arab-Americans, for example, is complicated by the complexity of both Arab and American, the way in which both reflect diverse experiences and identities. The same can be said for the term black-British. Current discussion of the movement of people (legally and illegally, voluntary and forced) across the globe has, on the one hand, increased the hybridity of people's collective identity. On the other hand, some people have reacted to such movement by reinforcing a belief in maintaining the "purity" of national identity.

It is false to separate the "domestic" and the "foreign" in an understanding of geopolitics. The geopolitical actions of states, the way they interact with agents external to their boundaries, require the support, tacit or overt, of their populations. The ideology of nationalism provides a sense of loyalty to the state and the belief that security rests upon sovereignty and integrity of the territory to which a national group lays claim. A component of nationalist ideology is the promotion of gender roles that facilitate a militarized foreign policy. In this chapter we have seen the ideological "glue" that maintains states and their geopolitical codes. In the next chapter, we explore a geographic feature that is also essential in maintaining the integrity of states and their national identities: boundaries.

Having read this chapter you will be able to:

- understand the connection between national identity and the state;
- identify the manifestations of nationalism in current affairs;
- identify the way gender roles are defined by the practice of nationalism;
- identify the important role of "combat" in the creation of national identities;
- understand how geopolitical codes are rooted in national histories.

Further reading

Cox, K.R. (2002) *Political Geography: Territory, State, and Society*, Oxford: Blackwell Publishers.

Parts of this textbook provide useful explorations of the state, including the local state.

Dahlman, C. (2005) "Geographies of Genocide and Ethnic Cleansing: The Lessons of Bosnia-Herzegovina" in Flint, C. (ed.) *The Geography of War and Peace*, Oxford: Oxford University Press, pp. 174–97.

This chapter explores questions that define a new field of inquiry, the geography of genocide.

Dijkink, G. (1996) *National Identity and Geopolitical Visions: Maps of Pride and Pain*, New York: Routledge.

Makes connections between the national identity and foreign policy using a number of extended case studies.

Dowler, L. (2005) "Amazonian Landscapes: Gender, War, and Historical Repetition" in Flint, C. (ed.) *The Geography of War and Peace*, Oxford: Oxford University Press, pp. 133–48.

An example of Dowler's work examining the connections between nationalism, gender, and conflict.

Enloe, C. (1990) *Bananas, Beaches, and Bases: Making Feminist Sense of International Politics*, Berkeley, CA: University of California Press.

Enloe, C. (2004) *The Curious Feminist: Searching for Women in a New Age of Empire*, Berkeley, CA: University of California Press.

Cynthia Enloe is the preeminent feminist scholar of militarism and militarization.

Murphy, A.B. (2005) "Territorial Ideology and Interstate Conflict: Comparative Considerations" in Flint, C. (ed.) *The Geography of War and Peace*, Oxford: Oxford University Press, pp. 280–96.

This essay was summarized in this chapter and provides the basis for the exercise in Box 5.3.

The following readings provide further exploration of nationalism:

Anderson, B. (1991) *Imagined Communities, Second Edition*, London: Verso.

Billig, M. (1995) *Banal Nationalism*, London: Sage Publications.

Smith, A.D. (1993) *National Identity*, Reno, NV: University of Nevada Press.

Smith, A.D. (1995) *Nations and Nationalism in a Global Era*, Cambridge: Polity Press.

The following readings provide further information on the conflict in Chechnya:

BBC News (2004) "Timeline: Chechnya," June 30, 2004. www.news.bbc.co.uk/2/hi/europe/country_profiles/2357267.stm. Accessed July 2, 2004.

Evangelista, M. (2002) *The Chechen Wars: Will Russia Go the Way of the Soviet Union?*, Washington, DC: Brookings Institution Press.

German, T.C. (2003) *Russia's Chechen War*, London: Routledge.

Knezys, S. and Sedlickas, R. (1999) *The War in Chechnya*, College Station, TX: Texas A&M University Press.

Politkovskaya, A. (2003) *A Small Corner of Hell: Dispatches from Chechnya*, Chicago, IL: The University of Chicago Press.

BOUNDARY GEOPOLITICS: SHAKY FOUNDATIONS OF THE WORLD POLITICAL MAP?

In this chapter we will:

- gain an understanding of the role boundaries play in geopolitics;
- define boundaries, borders, borderlands, and frontiers;
- discuss the role of boundaries in the construction of national identity;
- identify the boundary conflicts that most commonly appear in geopolitical codes;
- discuss how peaceful boundaries may be constructed;
- discuss the concept of the borderland and its implications for boundaries, nations, and states;
- provide case studies of the Israel–Palestine conflict and the Korean peninsula.

The boundary remains a material and ideological geopolitical feature. Despite eye-catching, or perhaps more accurately "book-selling," cries of the end of the nation-state and a borderless world, movement of goods and people (but less so ideas) is constrained by physical controls imposed by governments. Much of the geographic work on the porosity of borders and boundaries has been by European geographers looking at the internal boundaries of the EU. Alternatively, the War on Terrorism has promoted fears of "porous borders," especially the US's own, plus its policing of the Afghanistan–Pakistan boundary and that of Iraq. Similarly, as the boundaries of the EU are relocated eastwards, the public pressure on European government to focus attention upon refugees and other immigrants increases. In sum, the geopolitics of borders and boundaries remains, but the geography is the product of strong imposition on the one hand, and greater porosity on the other. See Donnan and Wilson (1999) for an excellent discussion of boundaries and borders, as well as the collection of essays on specific boundary conflicts in Schofield *et al.* (2002). In this chapter we focus on boundaries, but must note that their function is to control flow or movement. In the following chapter we concentrate on the geography of networks and the flows they facilitate.

First, we will define our terms and examine the ways in which boundaries are created and maintained, and the geopolitical role they play. To help understand contemporary conflicts we will provide a brief catalog of potential border disputes by examining the woes facing the fictional country of Hypothetica. We will then exemplify our discussion with a case study of the Israel–Palestine conflict. Boundaries and borders are also geographical features that may also reflect movements toward peace, the topic of the following section of the chapter. Finally, we will relate the demarcation of a boundary to global geopolitical conflict with a case study of the Korean peninsula.

Definitions

As with other topics, we will start by making sure we are using the same language, and we will adopt Prescott's (1987) terminology. The term *boundary* will be used to refer to the dividing line between political entities: the "line in the sand" if you wish that means you are in, say, Mexico if you stand on one side and the US if you hop over and stand on the other. Later we will look at the geopolitics of defining the precise location of the boundary and its effectiveness and role in controlling movement. The term border is often used synonymously with the term boundary, but for our discussion it is useful to distinguish the two. *Border* refers to that region contiguous with the boundary, a region within which society and the landscape are altered by the presence of the boundary. When considering neighboring states, the two borders either side of the boundary can be viewed as one *borderland*. This is especially useful when looking at the cross-boundary interaction between two states.

Finally, a term that is often used in the media when talking about boundaries is *frontier*. To be precise, a frontier refers to the process of territorial expansion in what are deemed, usually falsely, as "empty" areas. For example, the American frontier involved the killing, expulsion, and confinement of Native Americans to facilitate the land's "settlement" and its integration into the US economy. Even when indigenous populations were recognized, the creation of a frontier was justified through the language of religion and civilization: the regional population was a void for Christian practices to fill and integrate into the Christian realm. Echoes of this language remain today, as failed states are identified as the repositories of "evil" and, hence, must be brought back into the international state system and its norms of behavior.

Modern geopolitics was the politics of boundary construction. The building block of geopolitics was the nation-state, a political geographic entity that required territorial specificity as the basis for its sovereignty. Boundaries delineated the population and resources that came under the control of particular states. The geopolitics of mapping modern boundaries has three stages (Glassner and Fahrer, 2004). First, the course of the boundary must be *established*. This decision can be made through war, mutual political agreement, or external imposition. For example, we will see the role of external states creating the boundary between North and South Korea in a case study later in this chapter. The political boundaries in the continent of Africa are overwhelmingly the result of decisions made by colonial powers (Herbst, 2000). Once the boundary has been established it must be *demarcated*, its course must be made visible. In some cases, the visibility may not be clear on the actual landscape, but is solely a feature of maps. One could walk across the boundary without knowing it. In some cases the visibility of the demarcation is sporadic; checkpoints exist at trans-boundary roads and railways, but a fence does not extend along the full extent of the boundary. In extreme cases, the demarcation of the boundary is a violent expression, a continuous barrier of concrete, razor-wire, land-mines, attack dogs, and trip-activated machine guns (Figure 6.1). Not surprisingly, the form of demarcation is related to the degree of *control*, the third and final component of mapping boundaries (Figure 6.2). Decisions about the nature and intensity of flows across a border display great variation. North Korea is the most "closed" of all the contemporary states: goods, people, and information rarely travel out, and the opposite flow is sparse and completely controlled by the government. In the United Kingdom, entrance from other EU countries is relatively free, but there are many restrictions, made as visible by the government as possible for political capital, on refugees. The degree of control also varies with time; post 9/11 travelers entering the US have come under much more rigorous inspection, and required documentation has increased.

With so much effort being put into the establishment, demarcation, and control of boundaries, one must reflect upon the geopolitical purposes that boundaries serve. Within the geopolitical context of the War on Terrorism, boundary control is related to "security." States maintain their legitimacy, in part, by keeping their citizens safe, and control of borders is a pivotal factor. For example in the US the Office of Homeland Security was established in the wake of the terrorist attacks of September 2001 in order to enhance boundary security. Another example is Israel's success in establishing its boundaries in its quest to provide a territorial haven for Jews in a policy of Zionism.

The connection between boundaries and security is more complex than the ability to prevent invasion or infiltration. National identity is a territorial identity that rests upon

Figure 6.1 Closed border: Egypt–Israel.

Figure 6.2 Open border: Russian Caucasus.

the existence of, or desire for, a state with sovereignty over a piece of territory. National homeland, mythologized as it is, and state authority both rest upon territorial demarcation; boundaries demarcate nations and states and so define nation-states. Boundaries are, simultaneously, instruments of state policy, the expression and means of government power, and markers of national identity (Anderson, 1996). Their role in providing security extends into the taken for granted nature of national identity and citizen's expectations of government services.

The converse becomes of interest in discussions of the porosity of boundaries. If boundary control is, at least in some regions of the world, increasingly beyond the control of states, then what are the implications for national identity and state authority? We will address this geopolitical development later in our discussion of borderlands.

Geopolitical codes and boundary conflicts

Boundary conflicts remain a key motivation for states to go to war or make threats to do so. Figure 6.3 shows the sorry situation of a fictional country Hypothetica: a country that suffers from most of the usual grievances over boundary issues that can ignite conflict (Haggett, 1979). The separate issues can be grouped into four main categories: identity; control of natural resources; uncertainty over demarcation; and security.

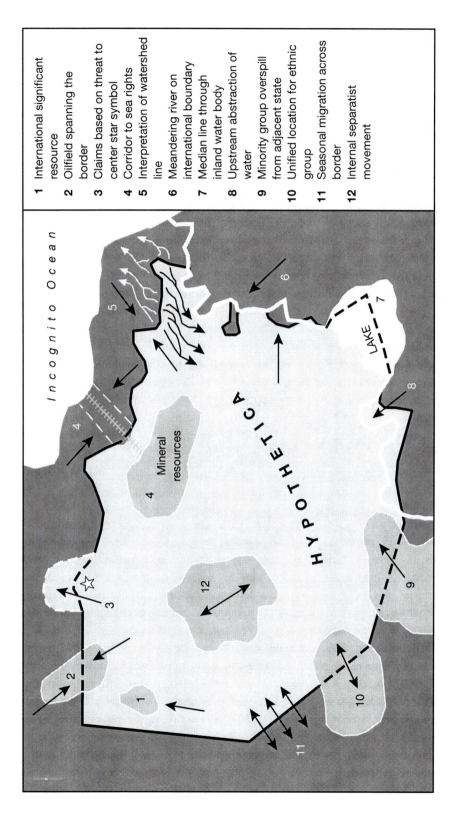

Figure 6.3 Hypothetica.

Legend:

1 International significant resource
2 Oilfield spanning the border
3 Claims based on threat to center star symbol
4 Corridor to sea rights
5 Interpretation of watershed line
6 Meandering river on international boundary
7 Median line through inland water body
8 Upstream abstraction of water
9 Minority group overspill from adjacent state
10 Unified location for ethnic group
11 Seasonal migration across border
12 Internal separatist movement

Incognito Ocean

HYPOTHETICA

Mineral resources

LAKE

Identity

In discussing our definitions and functions of boundaries, we saw that they play an important role in the geopolitics of nationalism. Nations require or desire the establishment of boundaries; they provide the legitimacy and power of the state. The geopolitics of an internal separatist movement reflects a perception that a group within Hypothetica has identified itself as a nation separate and different from Hypotheticans. For the separatists, the boundaries of Hypothetica do not provide a meaningful territorial marker for their national identity, and the boundary needs to be redrawn so that a new nation-state is created. The geopolitics of such a boundary dispute are likely to be difficult to resolve and the potential for violence is high, because the separatists' attempt to define national boundaries is an attack upon the notion of the territorial integrity of Hypothetica, an integrity that is the basis for its state power and national identity. The geography of the dispute also heightens the difficulties. The location of the separatists wholly within Hypothetica disrupts two related understandings of nation-states: a common nationality within the state's boundaries, and the territorial integrity of the nation-state.

The same issues exist for Hypothetica in two other locations. An ethnic group, with a collective identity distinct from both Hypotheticans and their neighbors, straddles the boundary. The primary collective allegiance of the ethnic group is not Hypothetican or the national identity of its neighbor. Perhaps the establishment and demarcation of the boundary ignored the location of this ethnic group, or decided that it was insignificant. On the other hand, the ethnic group may only have mobilized its identity into a political issue once the boundary had been established, and the control of the boundary prevented interaction between members of the ethnic group, patterns of interaction that were likely to have been established in the group's culture.

A similar problem exists to the southwest. The imposed boundary of Hypothetica transects historically established patterns of seasonal migration of pastoral peoples, following a path determined by the changing seasons and physical landscape in the search for water and fodder for their herds. The boundary does not take into consideration the functional needs of the pastoral peoples, their seasonal movement (or flow), possibly seen as "primitive," runs counter to the modern definition of nation-state spaces. In some instances, states may be unable to control such flows, or deem the seasonal movement as unimportant to national security. In other cases, the control of the movement may heighten as the geopolitical context changes, disrupting the social geography of the pastoral group.

The final boundary issue related to identity facing Hypothetica is a matter of the boundary's imprecise reflection of the geography of national identity. A minority group within Hypothetica has been created: a group that identifies with the national identity of the neighboring state. Political campaigns to unite such groups with the neighbouring national body are known as irredentism. As we saw in the discussion of nations and states, such situations may result in pressures by Hypothetica to expel the minority group and/or attempts by the neighboring state to redraw the boundary and capture some of Hypothetica's territory so that the minority is no longer outside the boundaries of "its" nation-state.

Demarcation

Demarcation of a boundary often reflects the physical geography of the landscape. Indeed, as we discussed in the typology of national myths, physical coherence may be the ideological basis of the nation-state (p. 125). The physical barrier imposed by mountain ranges has led them to be used as the basis for political boundaries, but this can result in an imprecise and disputed boundary demarcation. Logically, if a mountain range is to act as a boundary then the "center" of the range should be pinpointed. The physical center of the range is the watershed line, the line that divides the process of precipitation run-off; in other words, if a raindrop falls on one side of this physical line it would flow, say, east, but if it landed the other side it would flow west. In theory, this physical feature is definite and precise. In practice, especially in remote and rugged terrain, it is hard to define and demarcate across the whole extent of the mountain range or political boundary. Uncertainty in the course of the watershed line can result in different interpretations of the course of the boundary, resulting in conflicts regarding demarcation.

Another physical feature often used to demarcate boundaries is a river, often the thalweg or deepest channel of the river is used to pinpoint the course of the boundary. However, rivers are highly dynamic physical features. The flow of the water through the landscape creates erosion of the river's banks, and the course of the river will change over time. If the river has been used to demarcate a boundary, does the political boundary follow the old or new course of the river? If the old course of the river remains the official line of the boundary, what practical problems regarding fishing, agriculture, and water rights, for example, will emerge?

The final issue relating to physical features and boundary demarcation involves the use of lakes. If the boundary between states cuts through a lake, the norm is to define the median line between the shores as the boundary's line. However, erosion and changing water levels can provoke conflicts over the lines course, and the inability to paint a line on the water can lead to problems of control; precisely where does one state's jurisdiction end and the other's begin?

Resources

Boundaries define the territorial extent of a state's sovereignty, and sovereignty includes the right to extract and use resources. The course of a political boundary decides which states have access to which resources, and which states do not. Three resource related boundary issues are facing the sad and troubled Hypothetica. First, on the southern border, water resources are a concern. The neighboring state is upstream, meaning the land in that state is higher in altitude and the water travels through it before reaching Hypothetica. The water in the river is available for use and misuse before it crosses the boundary and reaches Hypothetica. The upstream state could, for example, use all the water in the river for irrigation or industry leaving the river dry and denying Hypothetica use of the water. Also, the upstream state could pollute the river, not only denying Hypothetica use of the resource but delivering it a problem of toxic waste and environmental risks.

In the northwest of Hypothetica an oilfield spans the boundary. Who has access to the oil, and, more specifically, how should the quantity of oil in the reserve be divided between the states? Next to the oilfield is a deposit of a particularly significant resource, uranium for example. Given the importance of this resource to the rest of the world, Hypothetica may face pressures to extract and sell the resource in a particular way. For example, uranium, essential for making nuclear weapons, is a resource that lies beyond the control of the state in which it is located. International agreements controlling how much, to whom, and for what purpose uranium can be sold, reduces the effective sovereignty the country has over its ability to sell the resource.

Security

The final set of boundary issues facing Hypothetica, a country I strongly recommend you do not invest your life savings in, fall under a general title of security. Hypothetica is a land-locked state, and so depends upon the goodwill of its neighbors to import and export goods by land. Particularly, the transport of mineral resources requires access to the sea, and so Hypothetica may negotiate for a territorial corridor to the ocean. Conflict can result if the corridor is not granted, controlled, in the eyes of Hypothetica, too rigorously, or closed once established. Finally, in light of potential or actual conflict with its northern neighbour, Hypothetica has invaded and now controls some of the land of its northern neighbour: the justification being that rocket or guerrilla attacks on town "A" were emanating from across the boundary, as in the case of Israel and its boundary with Lebanon and the Gaza Strip.

Activity

Look through an atlas of contemporary conflicts, such as Andrew Boyd's *Atlas of World Affairs* (1998), and see if you can relate the boundary conflicts identified in Hypothetica to real-world conflicts.

- In what way do the different types of boundary conflicts interact?
- Also, by looking at one conflict in detail think about how different social groups (class, race, gender, state bureaucrats, etc.) have different roles in these conflicts.

Case study 6.1: Israel–Palestine

The boundary of the state of Israel is often in the news, especially with regard to the Palestinian Authority, though low-level conflict continues with Syria and Lebanon too. Perhaps more than any other contemporary geopolitical issue, the conflict between Israel and the Palestinians is fought with "facts" as well as tanks and thrown stones. The history of the dispute is contested by each side in order to portray their current actions as just.

Here, I will try and give a "bare bones" history of the dispute in order to help us understand contemporary developments. I am sure it will not be to the satisfaction of anybody deeply committed to either side, but that is not its goal. I merely hope to provide some background to allow a reader who does not have a deep knowledge of the conflict to interpret media reports and also begin their own exploration of its causes, claims, and counter-claims. For a more in-depth discussion there are numerous sources, and each one will be perceived as biased. Well, they are. Here is my pick: Shlaim's *The Iron Wall* and *War and Peace in the Middle East* (2001), Friedman's *From Beirut to Jerusalem* (1995), Bregman and El-Tahri's *Israel and the Arabs* (2000), Drysdale and Blake's *The Middle East and North Africa* (1985), and Mansfield's *The Arabs* (1992). Also, to weigh the opposing views compare Said's *From Oslo to Iraq and the Roadmap* (2004) with Netanyahu's *A Durable Peace* (2000). The following history is my attempt to use these sources, and some additions, to highlight the historical basis of the contemporary conflict.

Similar to many boundary conflicts and related nationalist struggles, the Israel–Palestine conflict began with the dissolution of an empire. The Ottoman Empire was first established in the mid-1400s and at its peak had extended into Europe, across southwest Asia, and parts of north Africa. However, by the end of the nineteenth century it was in terminal decline. On the one hand, this resulted in some of its subjects seeking greater autonomy and independence. On the other hand, powerful countries such as France and Great Britain were extending their influence into Ottoman territory, as we saw in Chapter 4 with regard to Great Britain and Kuwait. The decline of Ottoman power was provoking both internal and external interest in establishing boundaries in territory that had been or still was under the declining control of the empire.

At the same time, the ideology of Zionism was gaining momentum. Zionism was creating a sense of Jewish national identity. It was a secular nationalism, with elements of socialist ideology, and its tenets were captured in Theodor Herzl's *The Jews State* (1896). From our earlier discussion of nationalism we know that in all nationalist movements a necessary connection between nation, state, and territory is made. Though other parts of the world were floated as possible sites for a Jewish state, the main focus was upon the biblical lands of Israel. The convening of the First Zionist Congress in 1897 encouraged and promoted Jewish migration to Palestine in a policy that was defined as, "a people without a land for a land without people." This statement is the kernel of the current conflict, for at the time more than 400,000 Palestinian Arabs lived in Palestine. However, within 30 years, Jewish immigrants outnumbered the Palestinian Arabs.

In World War I, the Ottoman Empire was one of the axis powers and the region was a key strategic theater. Though the Allies had defeat of the Axis powers in mind, there was also considerable rivalry and scheming on the side of the Allies. France and Britain used the war to jockey for position in a struggle between the two of them for greater control in the region after the war. There was much duplicity and tension, between the French and British, and between the two European powers and the Arabs with whom they tried to foster alliances. During the war, in 1917, British Foreign Secretary Arthur Balfour, in what became known as the Balfour Declaration, stated:

His Majesty's Government view with favour the establishment in Palestine of a national home for the Jewish people, and will use their best endeavours to facilitate the achievement of this object, it being clearly understood that nothing shall be done which may prejudice the civil and religious rights of existing non-Jewish communities in Palestine, or the rights and political status enjoyed by Jews in any other country.

(Fromkin, 1989, p. 297)

The two halves of the statement contradicted each other, as there was no plan on how a Jewish state could be established without compromising the existing Arab residents. In the wake of the Balfour Declaration and continued Jewish immigration, Arab-Jewish violence began around 1919. In one incident in 1929, 59 Jews were killed in Hebron. The volume of Jewish immigration increased in conjunction with lobbying efforts by Zionist organizations on France, Britain, and the US. In what was interpreted as a pro-Zionist move, the British government appointed a Jew and Zionist, Herbert Samuel, as governor of Palestine, a territory it now controlled in the wake of World War I.

Partly as a result of these developments an Arab rebellion lasted from 1936–9 in which 5,000 Arabs were killed, some as a result of aerial bombing by the Royal Air Force. The initiation of World War II altered Britain's geopolitical calculations. Britain needed cooperation from the new Arab states and territories to secure the flow of oil and, more importantly at this time, to maintain a continual territorial link with British India. As a way of nurturing Arab support, the British decided to try and stop the flow of Jewish immigration to Palestine to a trickle.

Unsurprisingly, Britain's policy met with resistance from the Jewish migrants. The policy was especially hard to justify in light of the Holocaust, Hitler's persecution of the Jews. At the end of World War II British policy was in tatters. Promises had been made to wartime Arab allies to limit Jewish immigration, but the Holocaust had energized the moral argument of Jews for a state. The Zionist movement resorted to violence, defined as terrorism, resistance, or national liberation, depending upon the political vantage point. A turning point was the bombing of King David Hotel in Jerusalem in 1946, the headquarters of the British administration, in which 91 people died. The intensity of the campaign led Sir Alan Cunningham, senior British official in Palestine, to admit the "inability of the army to protect even themselves." The 100,000 British soldiers in Palestine were unable to control the 600,000 Jews living there. The threat of violence toward the soldiers was so great that the troops were ordered to stay within their compounds, unless they left in groups of four with an armed escort. In a defining moment, two British sergeants left their compound, were captured by terrorists of the Irgun group, and killed with their bodies displayed for public viewing. The British government and people had had enough and handed over the situation to the UN.

The UN drew up a partition plan in November 1947. Under the plan, a Jewish state would control 56 percent of the existing Palestinian mandate, and an Arab state would control 43 percent. The city of Jerusalem would be a UN-administered, internationalized zone. The plan left no-one happy. The Zionists were upset as the Jewish state would not cover the whole of Palestine, as per the Balfour Declaration. Arabs saw a grave injustice with Israel receiving 56 percent of the territory, when Jews accounted for just one-third of the total population and owned just 7 percent of the land. Despite some misgivings, the Zionists accepted the partition plan, which was a generous territorial award and led to the recognition of the state of Israel.

In 1948, in the wake of the British withdrawal and the presentation of the partition plan, war broke out as the contiguous Arab states, plus Iraq, invaded to negate the establishment of the state of Israel. The war was a victory for the Israeli's: they ended up with all of the area allotted to a Jewish state by the UN plan, plus half of that allotted to the Arab state. Jordanian forces, with the help of Iraq, held the "West Bank" of the Jordan river and Egyptian forces held the "Gaza Strip." About 700,000 Arabs became refugees in Gaza, Jordan, and Lebanon, and approximately 700,000 Jewish immigrants arrived in Israel over the following 12 months. Simply put, the war of 1948 created the "de facto" boundaries of the state of Israel. East Jerusalem was controlled by Jordanian forces, but Israel proclaimed it as its capital, a move that was not recognized internationally.

In the decades that followed, a series of wars demarcated and established Israel's boundary with its Arab neighbors. In 1956, Egypt "nationalized" the Suez Canal sparking an invasion by British, French, and Israeli troops who feared that Arab control of the canal would disrupt use by Western countries of this vital trade route. The occupation was short-lived however, as the troops were forced to withdraw under intense US displeasure.

Two other wars followed that were more of a success for Israel and established the continued military dominance of Israel in the region as well as Israel's current boundary. The Six Day War of 1967 began when Egypt moved its army up to the Israeli boundary and blockaded the Gulf of Aqaba. In response, Israel attacked Egypt, Jordan, and Syria and easily captured the West Bank, Gaza, the Sinai Peninsula, and the Golan Heights. After this violent re-demarcation of the Israeli boundary, the Arab states responded with what is known as the Yom Kippur War of 1973 when Egypt and Syria attacked Israel over the religious holiday. Despite initial successes given the element of surprise, the Syrian army was soon defeated and an Israeli counter-attack encircled the Egyptian army.

The Yom Kippur War was a turning point in the conflict, though not a decisive one. Defeat led the Arab countries to reconsider the benefits of the relationship they had established with the Soviet Union, with the region being a strategic focus of the Cold War. While the UN brokered a gradual Israeli withdrawal from the Sinai peninsula, President Anwar Sadat of Egypt turned away from the Soviet Union and began to explore peace with Israel; a very brave initiative for any Arab leader. The result was the 1978 "Camp David" peace agreement that ushered in massive and continuing US aid to Egypt and Israel, but established the first peace agreement between Israel and a neighbor.

Of central significance to the conflict and hopes of peace are the UN resolutions 242 and 338 passed after the 1967 and 1973 wars respectively. The key points of the resolutions are:

- Israeli withdrawal from the occupied territories, meaning the West Bank and Gaza;
- recognition of the state of Israel and an end to the state of conflict;
- the right of return for Palestinian refugees was left vague and open to competing interpretations.

These resolutions have been the basis for the Palestinians claims to the West Bank and Gaza Strip; what they see as the necessary territorial foundation for a Palestinian state. With, at least in theory, goals of their own nation-states living peacefully side-by-side the Israeli state and the Palestinian people have attempted peace negotiations. These negotiations have been sporadic. At times, expectations and hopes of peace have been high, but at other times the situation has been confrontational. Israel, an independent state with a large and sophisticated military has dominated the Palestinians in terms of the ability to create "facts on the ground." A codeword for putting its military and people where they want to, despite their illegality under international law, diplomatic protest, Palestinian stone-throwing, and civil disobedience in an *Intifada* or uprising, and terrorist attacks upon Israeli citizens by factions of the Palestine Liberation Organization (PLO) and Hamas.

Attempts at peace have followed three general rubrics. Land for peace, or the "two-state" solution, calls for Israel to comply with resolutions 242 and 338 and withdraw from the occupied territories of West Bank, Gaza Strip, and East Jerusalem. The withdrawal would, so the story goes, be the basis for Palestinian national self-determination and sovereignty. Though this is often portrayed as a huge success or victory for the Palestinian people, the historical timeframe we have adopted in this case study shows that such a move would be seen as an enormous compromise by the Palestinian people: they would gain control of just 23 percent of what they see as their historic homeland. Put the other way, Israel would control 77 percent of the land covered by the Palestinian mandate.

The belief exists that "land for peace" would lead to "comprehensive peace": in other words, once Palestinian people achieve national self-determination then the Arab states would recognize the state of Israel and make peace once and for all. Though countries such as Egypt and Jordan have made steps along this path, continuing conflict, at various levels, with Syria and Iran for example, suggest that the connection or path should not be taken for granted. The harshest "plan" that exists is the Israeli rhetoric of "peace for peace"; or a construction of the conflict as Israel's self-defense against an untrustworthy enemy that is not worthy of the title "negotiating partner." From the Palestinian perspective, such a stance is not only being cavalier with history but fails to acknowledge the level of violence committed against the Palestinians—deemed greatly disproportional to the Israeli's loss from terrorism (Falah, 2005).

Land for peace rests upon Palestinian control of the West Bank. But what does "control" of the West Bank mean? Negotiations have given some concrete basis to who would control what in the West Bank, though the points of negotiation are contested within both the Israeli and Palestinian camps. Under what are known as the Oslo II Agreements of 1995, the West Bank has been divided into three areas (Figure 6.4):

- Area A: controlled by the Palestinian Authority.
- Area B: Palestinian civil control and Israeli security control.
- Area C: Israeli authority.

Under closer scrutiny, such a division strongly favours Israel. Area A comprises just 3 percent of the West Bank, Area B 27 percent, and Area C an overwhelming 70 percent.

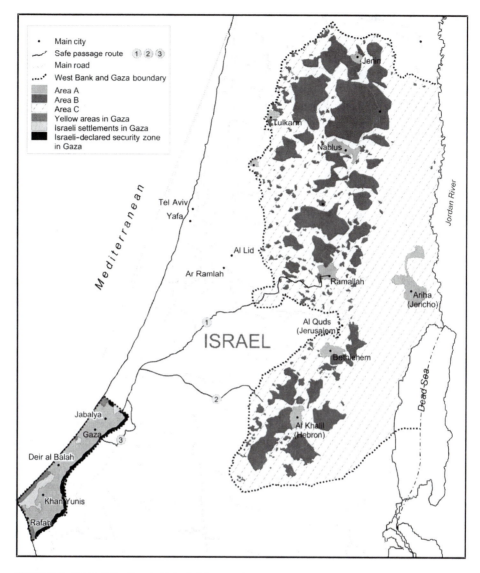

Figure 6.4 Israel–Palestine I: Oslo II Agreement.

In addition, Israel would maintain control of East Jerusalem. The geography of the division results in a "swiss cheese" state for the Palestinians, the small area they control being surrounded by territory under the control of the Israeli military (Falah, 2005). It would be like controlling Cardiff and Swansea, or Madison and Milwaukee, but not being able to move freely between them. The balance of power, or the element of territorial control, would remain firmly in the interests of Israel as they cite terrorist threats and overarching hostility to the state of Israel. As Prime Minister Sharon said, on January 28, 2003: "Palestine would be totally demilitarized ...; Israel will control all the entrances and exits and the air space above the state; Palestinians would be absolutely forbidden to form alliances with enemies of Israel" (Falah, 2005, p. 299).

The death of Yassar Arafat, long-term leader of the PLO and first chairman of the Palestinian Authority, was believed to offer opportunity for progress toward peace. The election of Mahmoud Abbas in January 2005 as the new president of the Palestinian Authority was met with hopes for peace, but also brinkmanship from Prime Minister Sharon who threatened renewed occupation of Gaza unless terrorist attacks were halted. Abbas had to satisfy both the Israeli's and Palestinian militants: a tough task that required the US and other influential countries to encourage talks and ensure that both sides act in good faith. In February 2006, Abbas was ousted by Hamas in free and fair elections.

However, while attention is often drawn to terrorist attacks by Palestinians, the Israeli government is using its dominant position to alter the geography of settlement and occupation that will (1) make the task of bringing militants into the political process very difficult, and may increasingly alienate the mainstream, and (2) creates "facts on the ground" that run counter to the spirit and goals of the UN resolutions, and (3) are violations of international law as well as violations of human rights.

Specifically, though attention has focused on Sharon's promise to remove Israeli settlers from Gaza (a promise fulfilled in August 2005), the process of establishing settlements in the more desirable and historically significant West Bank continues (Figure 6.5). International law prevents countries from building permanent structures and communities on land that they control through military occupation. Since the Oslo peace accords of 1993 and 1995, the amount of settlers on the West Bank has more than doubled to 200,000 including 30 new settlements. Of course, the different sides of the conflict will portray this settlement in different ways. On the one hand, the claim is made that Jewish settlements constitute only 1.7 percent of the land of the West Bank. However, when the full extent of the municipal boundaries is considered, as well as the territorial extent of the authority of Jewish regional councils then the coverage extends to 6.8 percent and 35.1 percent respectively (Falah, 2005).

The increase in settlements is paralleled, and some would say, facilitated, by the building of "the Wall" or security fence along a route that is based upon the "Green Line" boundary, but with some key exceptions (Newman, 2005). The Israeli government emphasizes that their security needs are being met by the construction of the Wall; it is seen as a barrier to prevent suicide bombers and other terrorists entering Israel and killing their citizens. There is, of course, some grounds for their stance. However, the Wall has been imposed upon the Palestinian population with no consultation and has amounted, in some cases, to a "land-grab", as some Palestinian villages have found themselves on the Israeli side of the Wall.

Finally, the issue of human rights remains when considering Israel's treatment of the Palestinian population. The destruction of Palestinian homes and olive groves that have stood for generations has been a constant focus of complaint by the Palestinians and human rights organizations. An Amnesty International (2004) report notes how in the three to four years leading up to 2004, 3,000 homes, and hundreds of public and commercial buildings, plus vast areas of agricultural land were destroyed in the Occupied Territories. It is estimated that around 20,000 people were displaced (Falah and Flint, 2004).

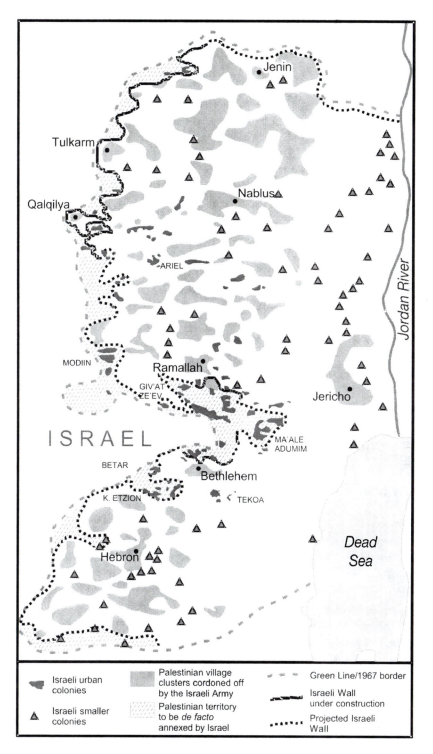

Figure 6.5 Israel–Palestine II: Palestinian villages and Israeli settlements.

What are the sticking points or major barriers to peace in this conflict? Four main issues stand out:

■   dispute over the control of Jerusalem. Palestinians have been increasingly excluded;
■   right of return for refugees;
■   a sovereign Palestinian State, but one that has territorial congruity and meaningful sovereignty from Israeli demands;
■   unequivocal recognition of Israel and its right to exist in peace with its neighbors.

The construction of the security wall, the Israeli settlement of the West Bank, the continuing Israeli control of Jerusalem, continued terrorist efforts by Palestinian militants, the dubious efficacy of the Palestinian Authority to harness the militants, unresolved boundary disputes with Syria, the political rise of Hamas and doubts about the US's ability to play the role of "honest broker" conspire to make the path to peace most challenging.

The case study illustrates some important points about the geopolitics of boundaries. Demarcation and establishment through war are unlikely to make a peaceful boundary. The political goodwill necessary for the construction of peaceful boundary relations is nigh impossible to cultivate when there is gross disparity of power between the geopolitical agents. Boundary conflicts are not merely the product of the local geopolitical codes of neighboring states but are often the product of the geopolitical codes of other states. Boundary disputes are inseparable from the politics of nationalism, and so identity plays a central role—including particular interpretations of history. Identity and control of movement were seen to be the key issues in the Israel–Palestine conflict, but these central issues may also be seen in an opposite light, as the sources for peaceful cross-boundary interaction.

The geopolitics of making peaceful boundaries

Boundaries are the focus for a variety of geopolitical disputes, but does the concentration upon the geographic line in the sand, an absolute marker of national identity and state sovereignty, provoke conflict? Boundaries create an absolute world of being either completely within a particular nation-state, or completely outside of it. There is no grey area in this geopolitical vision; the resource is either Hypothetica's or not, an individual is either a Hypothetican or not. Some argue that a more productive approach is to emphasize the geopolitics of borders rather than boundaries. Reflection upon borders and borderlands may result in trans-boundary interactions that allow for mutual control and utilization of resources and joint economic activities.

To make a peaceful boundary political goodwill between neighbors is fundamental: mutual trust and shared goals are the basis for cooperation (Newman, 2005, p. 336). Specifically, the following conditions are necessary to facilitate trans-boundary interaction (Newman, 2005, p. 337):

1   Territorial questions are settled. There is no dispute over where the boundary has been established and how it has been demarcated.

2   Trans-boundary interaction within the law is easy. The boundary facilitates flows (tourists and labour migrants, for example) between neighboring countries rather than preventing them.

3   The boundary provides a sense of security. Rather than being seen as a source of potential conflict, the boundary is seen as a sign of strength as commuting and joint economic projects enhance well-being and eradicate concerns of potential warfare.

4   Joint resource exploitation is possible. The basis of the peaceful boundary is mutual economic growth through interaction. For example, shared lakes, rivers, and aquifers may be managed jointly. Other examples are the "peace parks" or "free enterprise zones" that minimize the existence of the boundary by creating tariff-free international trade. The boundary as the enclosure of state imposed taxation is loosened by these zones.

5   Local administration is coordinated. Emergency services and transportation logistics are examples of how local governments in neighboring states can create functional integrated areas that straddle an international boundary.

In introducing trans-boundary cooperation, the focus was upon how two states interact politically for economic purposes. The coordination of local administration facilitates interaction, with the main goal being economic gain; increased trade, commuting to work across a political boundary, or jointly harvesting timber or fishing a lake, for example. The assumption is that the increased economic efficiency will strengthen the legitimacy of the separate states. However, cooperation may provoke other questions and concerns. What about issues of identity, if the role of the boundary in delimiting national identity diminishes, and what impact does this have on the way individuals in the borderland identify themselves?

Borderlands

Interest upon the cultural question of identity has focused attention upon borderlands (Martínez, 1994). The borderland is a trans-boundary region that shares common cultural traits, producing a geographic region of identity that is different from the two contiguous national identities. The borderland trans-boundary identity challenges the ideology that state boundaries encompass a national identity (Appadurai, 1991). Instead, borderlands require consideration, on the one hand, of the fractured nature of national identities, and, on the other hand, the commonalities (rather than differences) across national groups.

There are five key processes that shape a borderland (Martínez, 1994):

1   Transnationalism: borderlands are influenced by, and sometimes share the values, ideas, customs, and traditions of their counterparts across the boundary line. Hence, the ideological unity of national culture is challenged, as is the idea of state boundaries acting as the "containers" of national identity.

2   Otherness: the borderland is culturally different from the majority of both of the state's population of which it is part. The majority of the two state's populations

view the inhabitants of their border region and, perhaps to a lesser extent, the whole borderland as exhibiting different cultural traits.

3 Separateness: the cultural difference or otherness of the border and the borderland can result in an ideological and functional separateness from rest of state. Separateness may manifest itself in discrimination toward the border culture in education and the media, possibly with manifestation in government employment. In addition, the states' infrastructure may be relatively inefficient in the border region. Either alone, or in combination, cultural and functional separateness can make the two borders peripheral to their respective states. In light of this status, shared cultural traits across the boundary may foster solidarity and cooperation.

4 Areas of cultural accommodation: peripheral status and discrimination within their respective states may encourage the residents of a borderland to forge a sense of solidarity that transcends ethnic differences. The "them" and "us" dichotomy that a state boundary fosters can be undermined as collective identities that cross a state boundary and challenge national homogeneity are created.

5 Places of international accommodation: functional cooperation and cultural fusion can foster borderlands as zones of international cooperation, especially if economic integration and joint security and military operations have muddied the notion of state sovereignty being a singular enterprise that stops at the boundary. Instead, responsibility for security and economic growth is shared by two states, and its scope is no longer bounded by what has been understood as the geographical limits of the state.

The reason why scholars have increasingly focused upon borderlands is the role they play in creating geographies of identity and economic cooperation that are not based upon state boundaries and their ideological overlay with the pattern of national identity. If the boundary is key in establishing a state and nation, borderlands could play a role in challenging states and nations.

The geopolitics of identity, of which borderlands are one example, is challenging the importance of the hyphen in nation-state (Appadurai, 1991). The ideology of the nation-state asserts that all those within the boundaries of a state are members of a common nation. Going back to Chapter 5 on nations and nationalism, we saw that national separatist movements are practicing a geopolitics based on the idea that a particular state contains more than one national identity, and minority nations have a right to their own state. Appadurai alludes to a different geography: the geography of cultural groups is not a mosaic of nations that can be given territorial expressions as nation-states. Instead, cultural groups are tied together across the globe in networks of migration and cultural association that is played out over and in the boundaries of states. Networks of cultural association intersect state boundaries. Territorial manifestations of identity are sub-national connected to regions and localities within states. As ethnic groups settle in particular parts of a state they may construct a regional identity. Alternatively, the group may assimilate and move within the state, which reduces geographic concentration over time.

Case study 6.2: global geopolitical codes and the establishment of the North Korea–South Korea boundary

Korea's recorded history dates back to 57 BCE, dominated by periods of subservience to the Chinese Empire. However, this changed in dramatic form at the end of the Sino-Japanese war of 1894–5 when both Japan and China recognized Korea's complete independence. In the wake of Japan's victory, conflicting Japanese and Russian interests in Korea led to the Russo-Japanese war of 1904–5. Japan's victory stunned the Western world, where dominant racist ideology had made an Asian victory over a European state unthinkable. The final settlement to end the war was brokered with the aid of the president of the United States, Theodore Roosevelt. Japan was permitted to occupy Korea through the Treaty of Portsmouth of September 1905. By 1910, Korea was forcibly annexed and incorporated into the Japanese empire (Collins, 1969, p. 25).

Korea was administered as a colony within the Japanese Empire between August 22, 1910 and September 6, 1945. Facing both Chinese and Soviet attempts to exert influence in Northeast Asia, Japan became increasingly anxious to develop a regional geopolitical code. Korea was a key part of Japan's expansion into mainland Asia. In a quid pro quo between global and regional geopolitical codes, the United States and Britain were willing to give Japan free reign in Korea in exchange for Japanese recognition of their interests in Asia and the Pacific. Japan justified its occupation by portraying it as a "civilizing mission" of modernization (Hoare and Pares, 1999, p. 69). However, the objective of these developments was to turn Korea into a dependable and productive part of the Japanese Empire (Hoare and Pares, 1999, p. 69). Furthermore, the occupation was brutal, fostering an animosity toward Japan that remains, to some extent, today.

The animosity bred a nationalist geopolitical code of resistance. At the beginning of the twentieth century, camps were established to train a military force to resist the Japanese occupation, while other groups tried to gain assistance for the independence of Korea in a more diplomatic way, lobbying foreign governments. For example, Syngman Rhee, later to be the first president of South Korea, established the Korean National Association in Hawaii in 1909 (Eckert et al., 1990).

In the wake of World War I, the United States began to disseminate a global program of national self-determination. Koreans interpreted the context as one in which the major powers would be sympathetic to their own goals of ending the Japanese occupation. On March 1, 1919, a peaceful uprising burst out when a Declaration of Independence, prepared primarily by religious groups, was read out in Seoul. In the wake of fierce suppression, many Korean nationalists fled to China. A Korean provisional government was established in Shanghai in April 1919. However, the Korean exiles were very scattered and divided politically. These divisions were reflections of different perspectives on how to bring the Japanese domination of Korea to an end, as well as various ideologies (Hoare and Pares, 1999, p. 24). In other words, though the geopolitical goal of the groups was common, the means to achieve it were disputed.

The establishment of the Soviet Union had promoted the diffusion of social revolutionary thought. Socialism spread first among Korean exiles in the Russian Far East,

Siberia, and China, and then among Korean students in Japan, attracted by its combination of social change and national liberation. The different groups of exiles continued to clash, sometimes violently, over ideological differences. The Korean nationalist movement was too weak to end Japanese occupation. Instead, Japan was driven out of Korea in the wake of its defeat in World War II and the dissolution of its empire. Differences among Koreans remained unresolved (Hoare and Pares, 1999, p. 24).

Almost immediately efforts were made to form a Korean government with its headquarters in Seoul. Initially named the Committee for the Preparation of Korean Independence, on September 6, 1945 the government changed its name to the Korean People's Republic (Cumings, 1997, p. 185). Soviet troops had been fighting the Japanese in Korea since August 8, 1945. They gave "permission" for US troops to enter Korea further south than Seoul, while supporting the Korean People's Republic (Cumings, 1997, p. 186). As part of the redefinition of the US geopolitical code at the beginning of what came to be known as the Cold War, it did not recognize the republic the Soviet Union had helped create. In a move that presaged the division of Korea, the US chose instead to support the nationalist exiles and the few conservative politicians within Korea who comprised the Korean Democratic Party (KDP). Within a context of competition between two external powers, Koreans made political choices and within a matter of months Korea was divided into socialist and capitalist political allegiances with, virtually, a North and South geographic expression respectively (Cumings, 1997, p. 186).

The subsequent division of Korea had no historical or political basis. "If any East Asian country should have been divided it was Japan" writes Bruce Cumings (1997, p. 186), given its role as aggressor in World War II. For Koreans, the thirty-eighth parallel that was originally chosen to divide Korea had no prior meaning for Korean's, but now is central to their lives (Cumings, 1997, p. 186). Instead, the demarcation of the boundary was a product of the geopolitical codes of the Soviet Union and US. The thirty-eighth parallel was first established as the dividing line of Korea on August 10, 1945 by Dean Rusk and Charles H. Bonesteel, two American colonels who had been instructed to do so by John J. McCloy of the State-War-Navy Coordinating Committee (SWNCC) (Cumings, 1997, pp. 186–7). Their rationale was to include Seoul, the capital city, within the American Zone. Surprisingly, the Soviets accepted the division. Unbeknown to the Americans, the Soviets and the Japanese had themselves discussed dividing Korea into spheres of influence at the thirty-eighth parallel. Rusk confessed many years later that "[h]ad we known that, we all most surely would have chosen another line of demarcation" (Oberdorfer, 2001, p. 6). The decision was made without consulting any Koreans (Cumings, 1997, p. 187).

On August 15, 1948, the US-backed Republic of Korea was officially proclaimed and on September 9 the Soviet-backed Democratic People's Republic of Korea was proclaimed in the North. The Soviet Union chose Kim Il Sung (born Kim Song Ju), a 33-year-old Korean guerrilla commander who had initially fought the Japanese in China but had spent the last years of World War II in Manchurian training camps commanded by the Soviet army, to lead its regime in the North. In the South, the US chose 70-year-old Syngman Rhee as the first Korean president. He was a product of contacts with the US, and had obtained degrees from George Washington University, Harvard, and Princeton. Both leaders felt they were destined to reunite their country.

After the creation of these regimes both Soviet and US troops left the peninsula in 1948 and 1949, respectively. Just a matter of weeks after the US troop withdrawal, civil war broke out in the peninsula. On June 25, 1950, North Korea, with the support of the Soviet Union and China, invaded the South in an effort to reunify the country by force. The invasion was challenged and repulsed by the forces of the United States, South Korea, and 15 other states under the flag of the United Nations (Figure 6.6). The United States pledged support for South Korea against North Korea and sought legitimacy through the UN. In resolution 83 of June 27, 1950, the UN Security Council recommended that the member states of the UN should provide assistance to South Korea. The UN created a "unified command" (Hoare and Pares, 1999, p. 194), and asked the US to name a commander. President Harry S. Truman appointed General Douglas MacArthur as head of the unified command. MacArthur had also been the Supreme Commander for the Allied Powers in Japan (Hoare and Pares, 1999, p. 194). The distribution of ground forces for the UN Command was 50.3 percent US, 40.1 percent South Korean, and 9.6 percent others. The United States provided the majority of naval and air force units.

The invasion came after Kim Il Sung had repeatedly requested authorization from Joseph Stalin, the Soviet leader. Stalin eventually approved the war plan due to what he called the "changed international situation." What this means remains debated. Possible reasons are the victory of Mao's Communist Party in China, the development of the Soviet Union's atomic bomb, the withdrawal of US forces from South Korea, or a statement by Secretary of State Dean Acheson excluding South Korea from the US defense perimeter; all of which occurred in 1949 or early 1950 (Oberdorfer, 2001, p. 9). The Korean War was a proxy war, a war fought between the superpowers through their allies rather than direct conflict between the Soviet Union and the US. The war lasted from 1950 to 1953, and fortunes swung back and forth until an armistice agreement between the two Koreas was signed on July 27, 1953 (Hoare and Pares, 1999, pp. 3–4). The nature of the agreement means that the war is still unresolved; no final treaty has been signed. It is estimated that 900,000 Chinese and 520,000 North Korean soldiers were killed or wounded, as were 400,000 UN Command troops, nearly two-thirds of them South Koreans, 36,000 US soldiers were killed (Oberdorfer, 2001, pp. 9–10).

The end of the fighting resulted in the demarcation of a boundary close to the thirty-eighth parallel, a process initiated and defined by foreign countries. To this day the very limited flows across the boundary are controlled with the assistance of US soldiers stationed in South Korea, and the boundary is highly militarized. On the South Korean side, minefields line the roads, bridges are fortified, and checkpoints and gun emplacements are visible. The North Korean border is inaccessible. The war is, technically, still going on, and even today there are still fears in both Koreas that the fighting could break out at any moment.

In the aftermath of the fighting, the Rhee regime in the South became increasingly dictatorial and corrupt until it was deposed in 1960 by a student-led revolt. There were numerous coups and assassinations in South Korea until its government finally seemed to normalize in the late 1980s. In the North, Kim Il Sung systematically purged his

Figure 6.6 Korean peninsula.

political opponents, creating a highly centralized system that accorded him unlimited power and generated a formidable cult of personality (Oberdorfer, 2001, pp. 10–11). Kim Il Sung was in power for nearly five decades, and died of a heart attack in 1994, and was succeeded by his son Kim Jong Il.

In the late 1990s North Korea experienced terrible famines, killing approximately two million people and devastating life in North Korea (CNN, 1998). When Kim Dae Jung took office as president of South Korea in 1998, South Korea changed its geopolitical code toward North Korea from the hard-line policy of the Cold War to an engagement policy known as the "sunshine" policy (Heo and Hyun, 2003, p. 89). The US generally supported the "sunshine" policy during President Clinton's administration, and sought to negotiate an end to North Korea's development of nuclear weapons and long-range missiles (Heo and Hyun, 2003, p. 89). The US evaluated the politics of easing tensions over the Korean boundary as a means to advance its global geopolitical code. Since George W. Bush took office the overall picture between the three countries has changed, as the US defined a hard-line policy toward North Korea. President Bush included North Korea in his "axis of evil" defined in his 2002 State of the Union speech, and increasing focus has been placed upon North Korea's capabilities to build nuclear weapons and long-range missiles. Such statements by the US have hardened the attitude of North Korea and put the brakes on negotiations aiming to increase some flows of goods and people across the boundary.

The story of the Korean peninsula is one of a militarized boundary that is virtually closed to movement. The boundary is a product of external geopolitical influence that reached its most violent form to date in the Korean War. Its establishment, demarcation, and control were a component of the Cold War. More recent attempts by Koreans to change the boundary regime have been hindered within a new geopolitical context that has focused US attention upon North Korea's nuclear capabilities. To date, the nature of the Korean boundary is very much a product of geopolitics operating at the global scale, making any intermittent agency by the two Korean states toward a more open boundary problematic.

Boundaries and geopolitical codes

Boundaries and borders are an integral component of a state's geopolitical code. The legitimacy and tenure of a government depends upon its ability to maintain boundaries from external threat. The identity of a nation depends upon the effective use of the boundary in maintaining a sense of geopolitical "order" which is the maintenance of a particular domestic politics in the face of "outside" threats. The separation of a domestic "inside" from an "outside" realm of foreign policy has always been a fiction, but, arguably, this is increasingly so in the wake of intensified economic integration of the globe and related cultural and migratory flows. Nevertheless, governments feel the need to maintain the distinction in their policy and rhetoric. To take the British example again, the goal of introducing identity cards for all British citizens was built upon a geographic understanding of the world: the inside group had rights and privileges that needed to be protected from the undeserving "outsiders." However, such an "inside" and "outside" distinction denies Britain's historic and contemporary connections across the globe. The politics and demographics of the United Kingdom, as with other countries, are the product of colonial expeditions abroad that have resulted in dramatic social changes "back home." For example, increased government control of the economy in the first half of the twentieth century was a function of the two world wars, and patterns of immigration and citizenship through the twentieth century reflect the construction of empire and Commonwealth.

At the end of Chapter 3 we discussed the establishment of the US's War on Terrorism as a geopolitical code. The emphasis in that discussion was the creation of US influence beyond its own boundaries, or a global projection of power within the role of world leadership. However, despite its global role the government of the US must still maintain its legitimacy by defending its boundaries. The terrorist attacks of September 11, 2001 violated the boundaries of the US like never before; to maintain its legitimacy (and get re-elected) President George W. Bush's administration had to convince the American people the boundaries would be secure in the future. Part of the rhetoric of boundary defense appeared within the justification of the military invasion of Afghanistan, as shown in Chapter 3, but the notion of boundary security, the protection of sovereign territory, was prominent too.

For example, in the following quotes (taken from Flint, 2004) the defense of a national territory through increased control and monitoring of movements through the US boundary was emphasized. In the words of Tom Ridge, about to become secretary of the newly developed Department of Homeland Security at the time:

> It's one war, but there are two fronts. There's a battlefield outside this country and there's a war and a battlefield inside this country.
>
> (Governor Tom Ridge, October 22, 2001.
> Excerpts from news conference on anthrax in postal workers,
> www.nytimes.com/2001/10/22/national/22CND-EXCE.html
> —accessed January 10, 2002)

In other words, an attempt was being made to define two separate spheres of geopolitics, internal and external. An invigorated policing of the boundary was the geopolitical practice that was invoked to make this possible:

> I'd like to note that the INS has been and continues to be a very vital player in this war on terrorism, in this investigation, as well as the ongoing process of protecting the American people from what we see as the forces of evil.
>
> (Jim Ziglar, the commissioner of the Immigration and Naturalization
> Service, AG outlines foreign terrorist tracking task force, October 31,
> 2001, www.justice.gov/ag/speeches/2001/agcrisisremarks10_31.htm
> —accessed, January 12, 2002)

Put more simply:

> You know, we used to think of America the beautiful, fortress America, trusting
> America. And we find that perhaps we've trusted too much.
>   (NBC Nightly News Tom Brokaw interview with Tom Ridge,
>   October 16, 2001, www.msnbc.com/news/643696.asp.
>   —accessed January 12, 2002)

In the wake of the terrorist attacks of September 2001, the US faced a dilemma: it felt the need to intensify its global reach and influence in an increasingly militarized form at the same time that it felt the need to enhance the impervious nature of its own boundary (Flint, 2004). As its actions blurred a sense of division between domestic security and global reach, the boundary became an increasingly significant component of US geopolitical practice, rhetoric, and popular identity.

Summary and segue

Boundaries are the product and process of geopolitical agency. They are geographical features that are the manifestation of geopolitical actions, but they are also dynamic and contested geopolitical ideas and policies. A number of agents make boundaries the target of their geopolitical actions (governments, terrorists, nationalist groups) and boundaries are also the outcome of geopolitical processes operating at global, state, and sub-state scales. Actual and perceived boundaries, whether in existence or potentially established, provide the structure for geopolitical actions—whether it be the norms of international diplomacy or the terrorist actions of nationalist movements.

However, the emphasis upon flows in the academic discussions of borderlands or the policy imperatives of the Department of Homeland Security, require us to consider a very different geopolitics: networks and flows that cross political boundaries and so connect different places and territories. In the next chapter we explore the geopolitics of networks through a discussion of terrorism.

Having read this chapter you will be able to:

- understand how boundaries are an important part of the practice of geopolitics;
- identify the types of boundary conflicts within current affairs;
- understand why the establishment of boundaries is an important geopolitical practice;
- consider how geopolitical agency can undermine or change the roles boundaries play.

Further reading

Appadurai, A. (1991) "Global Ethnoscapes: Notes and Queries for a Transnational Anthropology" in Fox, R.G. (ed.) *Recapturing Anthropology*, Santa Fe, NM: School of American Research Press.

A provocative discussion of identity that emphasizes diasporas and multiple identities rather than nationalism.

Boyd, A. (1998) *An Atlas of World Affairs*, tenth edition, London: Routledge.

Though a little dated, this collection of short essays and maps is an extremely useful and accessible introduction to most of the world's conflicts.

Bregman, A. and El-Tahri, J. (2000) *Israel and the Arabs: An Eyewitness Account of War and Peace in the Middle East*, New York: TV Books.

It is practically impossible to recommend one book on any conflict, especially one as contested as this. But this book does an effective job of describing the main historic events in the conflict with the use of interesting interviews.

Donnan, H. and Wilson, T.M. (1999) *Borders: Frontiers of Identity, Nation and State*, Oxford: Berg.

An excellent survey and discussion on the literature addressing borders and boundaries.

Martínez, O.J. (1994) *Border People: Life and Society in the US-Mexico Borderlands*, Tucson, AZ: University of Arizona Press.

An in-depth study illustrating the nature of borderlands and their impact on boundaries.

Oberdorfer, D. (2001) *The Two Koreas*, Indianapolis, IN: Basic Books.

A highly interesting and accessible introduction to the Korean peninsula conflict.

GEOPOLITICAL METAGEOGRAPHIES: TERRORIST NETWORKS AND THE UNITED STATES' WAR ON TERRORISM

In this chapter we will:

- introduce the term metageography;
- discuss the geopolitics of globalization;
- discuss the geopolitics of defining terrorism;
- identify the changing geography of terrorism over the past 100 years;
- identify the geography of contemporary religiously motivated terrorism;
- define the metageography of terrorist networks;
- introduce the geography of the War on Terrorism;
- discuss the policy implications of the geography of terrorism and counter-terrorism;
- situate contemporary terrorism and the War on Terrorism within Modelski's model.

Terrorism has come to dominate the language and practice of geopolitics. For example, Homeland Security, weapons of mass destruction, the proposed democratization of Iraq and Afghanistan, the internal politics of NATO, and the relationships between the US and Russia are all framed by the identification of terrorism as the most important threat to be addressed in the geopolitical codes of the major powers. Cohen (2002) argues that the terrorist attacks of September 11, 2001 did not change the world, but only the way the US perceives the world. However, in its role as world leader, the way the US has changed its geopolitical code in the wake of those attacks has implications across the globe. Alternatively, perhaps the world has changed: the 9/11 attacks being the most shocking and visible manifestation of new forms of networked geopolitical power.

In this chapter we will focus upon the geopolitics of networks. First, we will discuss the term metageography and its connection to the geopolitics of globalization. Then we show the necessity of a geopolitical perspective in understanding the definition and history of terrorism, as well as the particular challenge of the contemporary forms of terrorism that are motivated by religious beliefs. The final sections of the chapter will address the metageography of networks that frame the "War on Terrorism" and make contemporary counter-terrorism activities particularly difficult. The US being both prime target of al-Qaeda and the chief protagonist of the "War on Terrorism" reflects the challenges faced by a world leader in decline, as we return to Modelski's model.

Geopolitical globalization: a new metageography

The world has changed since the time of the classic geopoliticians, and even since Modelski first proposed his model. We now live within an era of globalization, a term used to describe the global economic, political, and social connections that shape our world. The state-centric view of the classic geopoliticians has been replaced by a contemporary focus upon globalization, or a geography of networks that cross boundaries and are expressions of power that cannot be tied to particular national interests. The networks cannot be connected simply to the interests of a particular country in the same way as, for example, Ratzel's vision reflected German interests and Mackinder's British goals. Geopolitics is not just the calculation of countries trying to expand or protect their territory and define a political sphere of influence; it is also about countries, businesses, and political groups making connections across the globe.

Metageography refers to the "spatial structures through which people order their knowledge of the world" (Lewis and Wigen, 1997, p. ix; Beaverstock *et al.*, 2000). Modern geopolitics, within the dominant framework of Anglo-American geography has disseminated a metageography of the world as a mosaic of nation-states, despite the artificiality of these geographic units. For years, conflict between states was the focus of geopolitics: in other words, geopolitics was the sub-discipline that examined the power relations within the assumed metageography of nation-states. But the intensifying transnational networks of globalization are an emergent metageography in which flows of goods, money and people across boundaries makes banks, businesses, and groups of refugees, for example, important geopolitical actors. Political power is not just a matter of controlling territory, it is also a matter of controlling movement, or being able to construct networks to ones own advantage across political boundaries (Figure 7.1).

Let us contrast the geopolitics of globalization to the political vision of the classic geopoliticians. The economic concerns of, say, Mackinder and the German school were, for them, solvable through the exercise of political power by their own countries and by the extension of political boundaries. Countries were the most powerful geopolitical agents. In the era of globalization the geopolitical agency of countries has been limited as economic decisions must be made with reference to transnational economic organizations such as the IMF or WTO. Interest rates and currency values are set by the reactions of global markets and, in some cases, the IMF. Economic sovereignty is limited. In addition, the geopolitics of globalization has led to a dramatic increase in the number of geopolitical actors, especially NGOs and social movement (see Box 7.1).

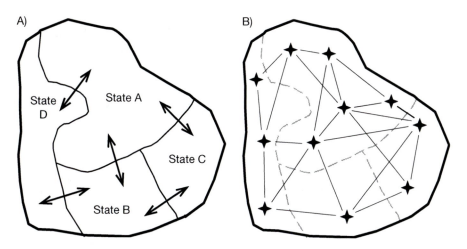

Figure 7.1 Metageography.

Globalization is the contemporary manifestation of what has been a constant trend in world history: the ever closer integration of parts of the globe. Technological improvements have allowed for quicker movement across longer distances for more and more people: from sailing ships through steam ships to jet passenger aircraft; from air mail through telephones to satellite technology; and from quite localized life experiences to global tourism and immigration. States have tried to manage this integration by, for example regulating domestic media markets to limit the number of outside broadcasts, and policing the flow of legal and illegal immigrants. However, the Internet and satellite TV have made it increasingly harder for states to manage the flow of information, and often the ability of states to manage international migration is limited.

On the other hand, it would be wrong to think of a simple dichotomy between states and networks. In many ways, states have been active agents in promoting transnational networks. Free trade and international investment is just one example of states negotiating to allow for the movement of goods and money across their boundaries. Increasingly, states are giving decision-making power to transnational organizations that have a direct impact upon the well-being of their population. For example, the WTO creates and adjudicates trade rules that have an impact upon jobs. The War on Terrorism has promoted a military network of cooperation between national police forces and armies across the globe (see Box 7.2).

The trend of increased integration is partially the product of the decisions and actions of countries. Of course, some countries are more powerful than others, and Modelski's framework suggests that the US, as world leader, should play a significant role. Indeed, the US has been active in securing freedom of movement for particular products and forms of investment, but has tried to protect American farmers and steel producers, for example, from cheaper imports. On the whole, the US has been active in promoting free trade and global investment. Plus, its dominance of the entertainment industry has facilitated the globalization of culture. Of course, other states have also been active in creating globalization. However, in Modelski's model we can view globalization as

Box 7.1 Geopolitics of resistance: the anti-globalization movement

The anti-globalization movement is a social movement, a collection of individuals seeking political and social change that operates outside of state institutions, that exemplifies the role of networks in contemporary politics. It has no territorial center or stable agenda, but is continually changing its methods and goals as a result of interaction between the diverse number of groups of which it is comprised. Reflecting this lack of hierarchy and its eclecticism the anti-globalization movement is also known as the Movement of Movements. The anti-globalization movement addresses a range of issues that range from ecological concerns to protests over economic neo-liberalism, to feminism. Such eclecticism produces no single and stable goal, leading to ridicule from those on the right of the political perspective, and criticism from those with a more traditional and state-centric left-wing agenda. However, its proponents claim that the fluidity of the movement is its very strength, enabling it to continually adjust to the dynamics of economic globalization and simultaneously showing the connections between issues of biodiversity, economic growth, democracy, and social marginalization. Furthermore, its lack of loyalty to a central organization prevents it from compromising on underlying beliefs; a multitude of movements will provide continual criticism, even of the movement itself. The number and diversity of movements creates connections across the globe to promote awareness of the way people in different places are connected by transnational economic and political networks. The movement has come together, though, in the World Social Forum conferences.

For more information regarding the content of past and future World Social Forums see www.nadir.org/nadir/initiativ/agp/free/wsf/ (accessed September 12, 2005).

being the outcome of economic and political decisions made directly by the US, or by other countries, within a political and economic rubric that has the world leader's blessing.

The cyclical nature of Modelski's model also suggests that there is another side to this argument. Some political geographers have argued that globalization has resulted in a new geopolitics, a new metageography, that has so undermined the power and sovereignty of states that no one state could attain the position of world leader again. Furthermore, others see globalization as a geopolitical "endgame": the last gasp of US global power that is another sign of its decline. In other words, by promoting globalization the US has sealed its own fate by transferring political and economic power into transnational networks in a way that the territorial base of state power that was the foundation of past world leaders is a matter of history and not contemporary geopolitical calculation.

To simplify the scenario, globalization may be seen as either the manifestation of US world leadership decline, with another cycle (American or otherwise) possible, or evidence of the end of modern state-centric geopolitics. Importantly, our understanding of structure and agency suggests that the future is not pre-determined. Geopolitical agents will continue to compete and one result will be the balance between the relative power of states and networks. The boxes in this section have exemplified two forms of

---

Box 7.2 Special Forces: the network power of the world leader

Networks of military power project the influence of the world leader across the globe. The increased role of US Special Forces since the invasion of Afghanistan in 2001 has been a mixture of covert military actions, but also "diplomatic" contacts with military forces across the globe. The former are military responses by the world leader to violent challenges, the latter are militarized attempts to maintain the US's global influence. The members of the Special Forces are highly trained and well-equipped soldiers, who have sought out the most dangerous form of modern combat. Ironically, much of their contemporary role consists of acting as "policeman," "diplomat," or, perhaps, "mayor" in conflict and post-conflict situations. Armed to the teeth, they are the visible expression of world leadership in the "hottest" conflict spots across the globe.

For example, since 1981, Special Forces Sergeant Rick Turcotte has trained Fijian forces for peace-keeping missions, operated covertly in the Honduran jungle to help US sponsored guerrillas in Nicaragua, and supervised military training in Thailand, the Philippines, Malaysia, Indonesia, and Singapore (Priest, 2003, p. 124). The training missions fall within

> the bread-and-butter mission of Army Special Forces . . . "foreign internal defense," a concept refined in successive campaigns against communism but yet to be fully adapted for the post-Cold War period. This task calls for special forces to "organize, train, advise, and assist" a foreign military so that it can "free and protect its society from subversion, lawlessness, and insurgency," according to Field manual 31–20, "Doctrine for Special Forces Operations," issued in April 1990.
>
> (Priest, 2003, pp. 128–9)

This quote contains clues to the interaction between the agency of the world leader and the metageographies of nation-states and networks. The definitions of "subversion, lawlessness and insurgency" are made within the world leader's geopolitical code. "Society" is used here as another term for state; it is particular countries that are being assisted. However, the assistance is provided through a network of military units that are under less political supervision, within the US and abroad, than regular units (Priest, 2003, p. 139).

networked power: social activism and the projection of military power. The remainder of the chapter will further exemplify the geopolitics of a network metageography by focusing upon the agency of terrorists.

Definitions of terrorism

The challenge to define terrorism is an impossible one for two reasons. First, terrorism has varied across history and geographical settings to make any one definition an inadequate description of the diversity of reasons and forms of terrorist activity (Crenshaw, 1981; Laqueur, 1987, pp. 149–50). Second, the definition of terrorism is in itself an act of politics: defining certain acts as terrorist acts makes certain forms of violence, political goals, and geopolitical agency illegitimate and so, in reverse, legitimates other forms of violence, politics, and agency. Defining a group as "terrorist" credits the form of violence that they inflict as being somehow "improper," "horrific," and "uncivilized." In calling these terms into question by no means condones the murder of people in the name of politics. Instead, the purpose is to think about how the category "terrorist" helps us to accept other forms of violence as "proper," "reasonable," and "civilized" (see Box 7.3).

Box 7.3 War crimes: power and *The Fog of War*

In the documentary *The Fog of War*, former US secretary of Defense Robert McNamara talks of his role as a strategist in the World War II firebombing of Japan that killed hundreds of thousands of civilians. In February 1945 one firebombing raid on the German city of Dresden destroyed 15 square kilometers of the inner city (Figure 7.2). Of the 28,410 houses in the area 24,861 were destroyed. Casualty estimates vary wildly, but recent scholarship puts the figure between 25,000 and 30,000, though some claim the total to be as high as 300,000. Overall, Anglo-American bombing of Germany in World War II killed approximately 400,000 people, about nine times the 43,000 British citizens killed by German raids. Japan also suffered firebombing. Beginning in February 1945, the four conurbations of Tokyo, Nagoya, Osaka, and Kobe were targeted. One attack on Tokyo in March destroyed 41 square kilometers and killed an estimated 100,000 people.

In the documentary interview *The Fog of War*, McNamara says that if the US had lost the war he would likely have been tried as a war criminal for his part in the bombing. Was the shared Axis and Allies policy of bombing towns in World War II an act of terrorism? Give an answer now, and reconsider it in light of the discussion of definitions of terrorism below.

Figure 7.2 Dresden after Allied bombing.

Undefined terrorism

In Bruce Hoffman's (1998) accessible introduction to the topic of terrorism, he takes great care to show the diversity of definitions of terrorism. Most telling is the table reproduced here (Table 7.1), which is a summary analysis of the predominance of particular terms or concepts in 109 definitions of terrorism (Hoffman, 1998, p. 40). I draw attention to this figure precisely because of the lack of agreement or consistency that it illustrates. The most agreed upon aspect of terrorism is violence, which appeared in just 84 percent of the definitions—in other words, 16 percent of the definitions did not emphasize violence as an important component of terrorism!

The definition of terrorism is, at best, contested and, perhaps more fairly, unclear. However, we can still discern some important geographical elements of terrorism from the features listed in Table 7.1. First, is the symbolic nature of terrorist actions that promotes the targeting of particular places or buildings. The Alfred P. Murrah Federal building in Oklahoma City was, for Timothy McVeigh and Terry Nichols, the local

Table 7.1 The problem of defining terrorism

	Definitional element	Frequency (%)
1	Violence, force	83.5
2	Political	65
3	Fear, terror emphasized	51
4	Threat	47
5	(Psychological) effects and (anticipated) reactions	41.5
6	Victim-target differentiation	37.5
7	Purposive, planned, systematic, organized action	32
8	Method of combat, strategy, tactic	30.5
9	Extranormality, in breach of accepted rules, without humanitarian constraints	30
10	Coercion, extortion, induction of compliance	28
11	Publicity aspect	21.5
12	Arbitrariness; impersonal, random character; indiscrimination	21
13	Civilians, noncombatants, neutrals, outsiders as victims	17.5
14	Intimidation	17
15	Innocence of victims emphasized	15.5
16	Group, movement, organization as perpetrator	14
17	Symbolic aspect, demonstration to others	13.5
18	Incalculability, unpredictability, unexpectedness of occurrence of violence	9
19	Clandestine, covert nature	9
20	Repetitiveness; serial or campaign character of violence	7
21	Criminal	6
22	Demands made on third parties	4

Source: Table is from Bruce Hoffman (1998, p. 40). Data source: Alex P. Schmid, Albert J. Jongman et al. (1988, pp. 5–6).

physical embodiment of the federal government that they viewed as an "occupying force" violating the freedoms of the American people. Less specifically, Palestinian terrorists target restaurants and busses in a brutal message that says that the public spaces of the state of Israel will never be safe until the rights of the Palestinian people for their own state are recognized (Falah and Flint, 2004).

Second, the goal of terrorism is to expand the geographic scope of a particular conflict in a manner that will, the terrorists hope, benefit their cause. Osama bin Laden, has made the presence of US troops on the Saudi peninsula a matter that we must all consider, and something that becomes a part of electoral campaigns in Australia, Spain, Great Britain, the US, and beyond. The terrorist's perceived need to reach a broader audience, or expand the scope of "interested" or at least "implicated" parties relates to the marginalization of some groups to the extent they resort to violence in order to place their situation on the political agenda. However, for marginalized groups to be heard, they must often change the scale at which their situation is discussed or decided: groups dominant in a particular state may well have no interest in hearing the complaints of the marginalized. Through acts of terrorism, marginalized groups may change the scope of the political debate, making it a regional or global issue, and so forcing the dominant group in the state to at least talk and maybe even address the situation.

Third, terrorist groups claim, in the words of Hoffman (1998, p. 43), to be performing political altruism. In other words, terrorists believe they are serving or speaking for a group who have been marginalized or oppressed and deserve a better political deal. A more exact understanding of the terrorist would be as a political geographic altruist. The motivation for terrorism is perceived political injustices, but these are insepar-able from particular geographic organizations of power relations (see Chapter 1 for a reminder). This is most clear in the case of terrorism motivated by nationalism; the goal is a reorganization of space to create a new independent nation-state. In the case of al-Qaeda, their motivation rests upon the marginalization of Arab influence in the world: specifically, for them, the violence meted out by Israel upon the Palestinians, the exploitation of oil reserves by Western companies, and the presence of US forces across the Arab world. The geographic problem is, broadly speaking, a "colonial" relationship that, it is argued, can be relieved by removing the US presence and eradi-cating the state of Israel. The motivation behind terrorism, and hence the possibility for lasting resolution, can only be fully understood through a recognition of the territorial expression of the politics at hand.

Though no single definition of terrorism is possible, the features of the definitions reflect the geography of the causes, and means of terrorism. Terrorism is an act of geopolitics that is motivated by the spatial manifestation of power, uses geography (in terms of symbolic places and expanding the scope of the conflict) in its tactics, and requires a rearrangement of existing political geographies if it is to be successful or peacefully resolved.

You're a terrorist . . . I'm not

In Chapter 4 we introduced the role of the representations of people, places, and states as an important part of geopolitics. Defining terrorism is also an act of representation that, by restricting the label "terrorist" to a few creates a wider set of actions and agents that are "non-terrorist." The key question in these acts of representation is the state: some definitions of terrorism are purposeful in emphasizing "non-state" or "sub-national" agents as those who commit terrorism, hence excluding the state as an agent of terrorism (Flint, 2005). Criticizing the omission of consideration of some state actions as terrorism does not imply that every state, throughout history, is a "terrorist." However, restricting terrorism to "sub-national" groups does prevent certain state actions at particular times being designated as acts of violence aimed at instilling fear in to the population for political reasons. Such state repression is usually undertaken to establish and maintain control by throttling political opposition. History would, it seems, allow for certain state actions to be seen as the use of violence to create a climate of fear and political compliance.

Adolf Hitler's actions in establishing Nazi Germany and Josef Stalin's political purges are seen as "classic" examples of the state becoming a "police state" to squash any political dissent and opposition. The early example of these states was continued as part of the domestic aspect of the geopolitical codes of states within the Cold War: from the McCarthy trials in the US in the 1950s that brought the power of the state judiciary to bear upon anyone proclaiming a left-wing political agenda and forced people to fear for their careers and reputations, to the secret police forces of the Communist regimes of Central and Eastern Europe. The geopolitics of the Cold War constructed domestic "threats" or "enemies within" who were hunted by the state often tortured and killed, one of the goals being to create a public atmosphere of fear that it was believed would prevent political opposition (see Box 7.4). Contemporary regimes in North Korea, Syria, and many others, some defined as "allies" in the US War on Terrorism, are guilty of the same actions for the same goals, to varying degrees.

Ahmad's (2000, pp. 94–100) definition of terrorism, purposefully constructed to allow for the inclusion of state actions, has another type of state violence in mind. Ahmad is referring to the actions of Israel, against the Palestinians, and India and Pakistan in the conflict over Kashmir. In these instances, the military wing of the state is using violence in a purposeful and systematic manner to quash nationalist movements that would alter the current boundaries of the state; and in the case of some of the rhetoric and interpretations of the Palestine-Israel conflict, the very existence of the state of Israel. Accusations, inquiries, and revelations still remain over the illegal use of force by the British government against the Irish Republican Army (IRA). When the territorial integrity of the state is challenged, the state may go beyond the realms of legality to counter national-separatism. In these situations, violence, diffusing fear through a wider population, and political goals (all common features of definitions of terrorism) are part of the calculations and actions of states. Terrorism? Finally, what of the deliberate and sustained bombing of civilian targets in World War II, as discussed earlier? The goal of these displays of military might was to sap civilian morale and cause surrender. Terrorism?

Box 7.4 The School of the Americas

During the Cold War the US established the innocuous sounding International Military Education and Training Program (IMET). The Program trained over 500,000 foreign officers and enlisted personnel. The main campus, the School of the Americas, was relocated to Fort Benning, Georgia in 1984. The title of the outfit illustrates that much of the program's regional focus was Central and South America. Defenders of the Program claim that it disseminated "American values" through trips to Disneyland and sporting events. However, the product of the School is far from the images of Disney. The School trained soldiers in "low intensity conflict." In other words, not how to fight an invading or hostile army, but to prevent counter-insurgency in some of the poorest and most polarized countries in the world ruled by undemocratic and brutal military regimes, such as Honduras, Haiti, Paraguay, Uruguay, Chile, Peru, Colombia, Panama, El Salvador, and Guatemala. The School of Americas includes a "Hall of Fame" displaying portraits of "successful" graduates. Infamous would be a more accurate description. To quote Chuck Call of the Washington Office on Latin America:

> In El Salvador, 48 of 69 people named in the UN Truth Commission Report as human rights violators, graduates of the school. Half of the people named in a recent report done by NGOs of alleged human rights violators in Columbia, 128 of 247, graduates of the School of Americas. This is at such a level that you can't ignore it. And what's important about that is that it associates the US military with these abusive forces.

Defenders of IMET admit a "few bad apples." Critics of the Program argue that the US trains torturers and killers targeting groups and people who support social reform.

The quotes and information in this box are from a video put out by the American Defense Monitor in 1994 entitled *School of the Americas: At War with Democracy?* The transcript is available at www.cdi.org/adm/804/transcript.html.

In what way does state sponsored torture and oppression fit the definition of terrorism, and in what way can it be argued to be something other than terrorism? How are your answers molded not by what is done but by who (a government agency) is doing it?

History of modern terrorism: waves of terrorism and their geography

In a useful, though necessarily simplified exercise, Rapoport (2001) has identified four separate but connected "waves" or periods of modern terrorism. Describing these waves offers not only a brief history of terrorism, but also highlights the changing geography of terrorism (Flint, 2005), a change that has important implications for the contemporary politics of the "War on Terrorism."

The goals and arena of the first two waves of terrorism were focused upon one particular geopolitical scale, the nation-state. The first wave occurred between, roughly, the 1880s and the beginning of World War I in 1914 and was motivated by the piecemeal political reforms of the Russian tsar hoping to preclude more radical and revolutionary change. The goal of the terrorists, loosely defined as "anarchists," was to mobilize the citizens of Russia toward revolution as they feared the population would be placated by the reforms: In other words, the terrorists wanted to change the way that the Russian state was governed. These "anarchist" politics diffused, with limited success, to other parts of Europe. The geography of this first wave was framed by an understanding that the state was the source of political change and so bounded the scope of action. Though the ideology of the terrorists, and the way they conducted terrorism, diffused from Russia into parts of Europe, the geography of the first wave of terrorism was restricted to within state boundaries.

To a lesser degree, the first wave of terrorism also reflected an increase in nationalist politics. The assassination of the Austrian Archduke Franz Ferdinand in Sarajevo by a nationalist sparked World War I, which in turn catalyzed many political and social changes. One of these changes was the explosion of demands for national self-determination, or the desire for people to create and belong to national communities synonymous with independent and sovereign states.

The second wave of terrorism (approximately 1920–60) was dominated by the political geography of ending imperialism, or decolonization, and the establishment of nation-states. Terrorism was, in some cases, deemed a necessary and useful strategy to force colonial powers to leave and, in a related politics, define which social and ethnic groups would play the key roles in defining the new state. Examples of this type of terrorism include the Irgun in Israel angry toward the British government's restrictions on Jewish in-migration, and the Mau-Mau in Kenya. The geography of this wave was similar to that of the first: the arena and goal of terrorism was the nation-state, in this case to establish a new one rather than change the politics of existing states. However, more so than the first wave, the impetus toward national self-determination was an agenda that spanned the globe. The US, as world leader, was promoting the dissolution of existing empires through the establishment of independent nation-states (Chapter 2).

The third wave of terrorism (1960s–90s) may also be interpreted as actions set within the context of the geopolitical structure of world leadership. The US's entrance into the conflict in Vietnam, and the growing dominance of the Vietnam War in political debate, was one ingredient in the growing influence of radical left-wing groups in Western politics. The Vietnam War fermented critique of the goals and actions of the world leader. Especially, its twin "innovations" development and national self-determination began to be questioned by left-wing political groups, as the escalation of the Vietnam War was interpreted as the old process of colonialism in a new guise. A component of this radicalism was the emergence of terrorist groups in Western Europe and the US motivated by Marxist ideology. For example, the Baader-Meinhof gang in Germany, the Red Brigade in Italy, and the Weather Underground in the US were all motivated by left-wing ideology. The political change that these groups professed was couched, to varying degrees, within a vision of "international Marxism," and the context for their actions was discontent with the global actions of the US, and its agenda as world leader.

Another side to the third wave of terrorism was nationalist groups who saw the project of decolonization and national self-determination as incomplete and unfair. Two prominent examples are the IRA and the Northern Ireland conflict and the PLO and its claims for a Palestinian state. The IRA had witnessed the decline of the British Empire across the globe, but called for the process to continue and allow for a united Ireland free of British rule. The PLO had witnessed the establishment of a new nation-state on the territory of Palestine, but it was the state of Israel. Nationalist conflicts remained a key motivation for terrorists in the third wave.

However, the geography of their actions was significantly different from the second wave. In the third wave a greater internationalization of terrorist activity became evident. Terrorist groups were still predominately based within particular states, and were focused upon change at the scale of the state, but they began to operate and cooperate across state boundaries. The PLO is a good example, using the tactic of hijacking international passenger flights to increase the geographical scope of its activity and generate an international audience for its political message. As airplanes run by British companies sat on the tarmac of foreign airports under the control of Palestinian terrorists and surrounded by non-British security forces, the issue of Palestinian self-determination became more than a problem for Israel and the Arabs. Perhaps the most poignant act was the 1972 Munich Olympic Games when Palestinian terrorists entered the Olympic village, a symbol of international respect and peace, and killed 11 Israeli athletes. Claims of the "whole world watching" were exactly the geographical outcome the terrorists were aiming for; the Palestine-Israel conflict became a matter of international importance and diplomacy.

The second form of internationalization in the third wave was the growing cooperation between terrorist groups based in, and identified with, different states. Training and weapons exchanges became a part of terrorism, and the networks of terrorism became an international rather than national phenomenon. Laqueur (1987) relates the internationalization of terrorism to the Cold War, and the growth in the 1970s of state sponsorship of groups originally defined by their territorial and nationalist demands. Internationalization was perceived by terrorist groups as a means of widening the scope of the conflict and hence increasing the "audience" for their cause. However, it also facilitated state-versus-state conflict. Various governments attempted to gain influence in a particular dispute by supporting different factions of the same cause such as Syria, Libya, Iraq, and other states funding separate Palestinian groups. The outcome of state sponsorship was to make terrorism "almost respectable," with a sufficient majority of states at the UN preventing any effective international coordination of counter-terrorist actions (Laqueur, 1987, p. 269). The Soviet Union and Libya were significant suppliers of weapons and funds to terrorist groups, but in the 1980s Syria and Iran became increasingly important (Laqueur, 1987, p. 295). Today, it is these latter two countries that are the focus of US statements on state-sponsored terrorism and the possible targets of sanctions or military force.

The third wave of terrorism may be interpreted within the global political structure identified by Modelski's model in Chapter 2 (see Table 7.2). The motivations, at least

Table 7.2 Waves of terrorism and Modelski's model

Wave		Terrorist groups	Geography	Interpretation within Modelski's model
I	1880–1914	Anarchists	Within states	Prelude to global war
II	1920–60	Nationalists	Within states Decolonization	Aggressive "pursuit" of world leader's innovation
III	1960–90	Nationalist Ideological	Internationalization	Delegitimation/ deconcentration
IV	1990–present	Religious	Transnational "cosmic"	Deconcentration?

in the ideology that the groups espoused, for terrorism were a reaction against two of the main components of the world leader's "innovation." First, the nationalist movements shared, if somewhat loosely, a belief that "imperialism" in the form of the domination of the rich countries over the poor remained. The rhetoric of the movements pointed to the Vietnam War as evidence, and in sum argued that the United States had reneged on its promises of national self-determination, a key element of its world leadership "innovation." Second, the Marxist framework for many of the terrorist groups in this period reflected an ideological and armed challenge to the "promises" of capitalism that the US used in the Cold War.

The fourth wave of terrorism (1990s–present) portends a much more dramatic geographical change with severe implications for both acts of terrorism and the effectiveness and implications of counter-terrorism. For Rapoport, the fourth wave of terrorism is the period of religious terrorism, though terrorism motivated by nationalism is far from gone. The geography of goals and beliefs of religious terrorists goes beyond international connections; it is a geography that "transcends the state," perhaps the state as political agent is irrelevant to this form of terrorism.

Christian, Jewish, Muslim, Sikh, and Buddhist religions are all tainted by groups who utilize a fundamentalist view of the belief system to justify acts of terrorism (Juergensmeyer, 2000). In other words, religious terrorism is a contemporary global phenomenon, and not limited to one particular religion, as politically motivated claims against Islam, especially, suggest. Religious terrorists are fighting a "cosmic war"; a war of good against evil in which the adjudicator is God or another form of supreme being, and the terrorists are merely the soldiers conducting God's will (Juergensmeyer, 2000). The battle, in the case of religious terrorism, is for people's souls and not a secular political agenda. The state may be the source of acts deemed "evil" but the state is not the answer, for that one has to turn to salvation and a different world.

Terrorism motivated by religious fundamentalism is a particularly dangerous form of terrorism. It is more likely to invoke terrorist acts that produce a large number of casualties and be less sympathetic to overtures of conflict resolution than the previous waves of terrorism (Juergensmeyer, 2000). Why? To understand this dreary prediction, we have to consider the way the state has dominated both geopolitical practice and analysis throughout the twentieth century. Geopolitical actors have seen the state to be the key structure that both constrains or motivates their actions, but it has also been seen as the key "prize": the geopolitical structure that, if controlled or changed, will reap political benefits. By waging a "cosmic war" religious terrorists have shattered this essential geopolitical assumption of the twentieth century, confounding policy-makers and academics in the process.

Religious terrorism, by fighting a "cosmic war" transcends the state as an arena for politics: the goal is to serve God's will and fight "evil". Essentially, the battle is of a spiritual nature and not secular. If that is the case, then victims are "infidels" or "sinners" whose death will, in the minds of the terrorists, please God. With these beliefs, religious terrorists do not need to make the political calculations of secular terrorists in which the number and type of casualties had to be balanced: enough to "shock" but not too many to alienate "sympathizers." For religious terrorists, their actions are part of one sort of Armageddon or another, and not the bloody side of a wider political process, and hence the lack of constraint on the number of casualties.

The second implication of the "cosmic war" thesis is that the state is no longer seen as the key geopolitical arbiter. The state as a structure that could enable terrorists and their sympathizers by providing political concessions, or even conceding defeat, is deemed irrelevant by religious terrorists. The question no longer becomes a matter of harassing politicians to address the terrorists concerns, as is usually the goal of nation-alist-separatist terrorists. Instead, the belief is that the state is the embodiment of the evil that, following God's will, needs to be destroyed. Again, restraint is not an issue, and the likelihood of large-scale horrifying attacks is increased. For example, Timothy McVeigh did not blow up the Murrah federal building, including the day-care centre, to bring representatives of the US government to the negotiating table. He killed what he saw as agents of evil destroying a "way of life" defined, if loosely, by religious beliefs. For religious terrorists, the state is an actor that needs to be destroyed and not negotiated with. The structure is spiritual and "cosmic," enabling acts of "martyrdom" beyond constraint if you perceive yourself to be acting on God's will.

But wait a minute. Does religious terrorism really "transcend" the state? There are good reasons to qualify such a claim. Two strands of argument can be made: religious terrorists still use or need states, and the goals of religious terrorism are still related to the state as the key geopolitical structure.

The identification of Afghanistan as the "home" or "base" of al-Qaeda immediately after the terrorist attacks of September 11, 2001 is testimony to the relationship between some terrorist groups using religion as their motivation and the need for the protection and sponsorship that can be offered by territorially sovereign states. In the next section we will discuss that relationship between terrorist networks and sovereign states at length. At the moment, it is enough to refer to bin Laden's relationships with the governments of Sudan and Afghanistan to see that this particular terrorist group actively sought the haven that territorial sovereignty can provide.

The second question is whether the goals of religiously motivated terrorism transcend the state. For example, interviews with Jewish settlers in the West Bank, and their recourse to scripture for motivation and justification make for compelling reading (Juergensmeyer, 2000). The belief that the land of Israel was "given" to the Jews by God is clearly part of the conscious that motivates the killing of both secular Jews and Arabs who are deemed to betray or threaten this "return" of Israel to the Jews. But what

of the goal? The goal is the establishment of state sovereignty across a particular territory, known as the West Bank or Judaea-Samaria depending on the perspective and
agenda. In the British Isles, the conflict in Northern Ireland is usually portrayed as a
nationalist struggle, yet Juergensmeyer (2000) emphasizes the religious vitriol between
the Protestant unionists and the Catholic republicans. Again, perhaps motivation is being
confused with goals. Both sides have agendas regarding the territorial extent of Irish
and British sovereignty.

The final issue in discussing whether religious terrorism transcends the states refers
to the role of state as arbiter in political disputes. The thesis of "cosmic war" rests upon
the terrorist's perception that God is judging their actions and will provide the subsequent rewards (Juergensmeyer, 2000). But, in some cases, the state has a role to play
in evaluating and delivering the terrorist's demands. This is most evident, perhaps, in
the case of the United States where the assassination of doctors performing legal abortions is the extreme manifestation of Christian right lobbying and protest to change the
laws of the land and ban abortions. With an increasing number of senators and representatives in Washington supporting a ban on abortion it is not inconceivable that access
to abortion will be restricted further and even banned. Whether this would be a "victory
for terrorism" is a matter of debate. The point is that if such a change in government
policy was to be legislated, the goals of terrorists motivated by Christian fundamentalism would have been achieved by the actions of the state.

In summary, terrorism motivated by religious beliefs does appear to be experiencing
a surge in activity across the globe and all the major religions. Religious terrorism is
creating geography different from those of the previous waves, as the state plays a less
central role. Resort to the scale of a "cosmic war" makes religious terrorists less chained
to the opportunities and constraints that exist when the state is seen as the key
geopolitical structure. This new geography of structure and agency has implications
for the severity of terrorist acts and the possibilities for conflict resolution. However,
the state is still an essential scale in the calculations of religious terrorists, whether as
a strategic territorial haven or the target of political goals. To understand religious
terrorism it is useful to think of two separate but closely related geographies: motivation is sought at the "cosmic scale" while goals and actions are still tied to the scale
of the state.

Resort to the "cosmic scale" may be explained within Modelski's cycles of world
leadership, because it is a rejection of the type of society promoted by the United States.
Religious fundamentalists reject many of the material and cultural aspects of the world
leader's culture as immoral: an attack on what are promoted as "traditional values."
Instead of wishing to adopt the world leader's innovations, religious fundamentalism
rejects them. However, it should also be noted that religious fundamentalists adopt
many aspects of modernity (especially weaponry, the Internet and other forms of
communication, as well as decentralized organization) (Armstrong, 2000).

The importance of religious terrorism in contemporary geopolitics has forced
policy-makers and academics to re-think the taken for granted understanding of geopolitics as inter-state politics. Hence, it requires us to focus upon terrorism and counter-
terrorism as involving two, perhaps incongruous, understandings of the world. So, we
turn to the metageographies of terrorism and counter-terrorism in the next section.

Activity

Return to bin Laden's *fatwa* described in Chapter 3 as a geopolitical code.

- Define the "spiritual" elements of the code.

- In what sense do they relate to Juergensmeyer's notion of a "cosmic war"?

- In what sense is the code focused upon territorial issues that can be interpreted through the political geography of state sovereignty?

- In what ways, if any, do the spiritual and territorial elements of the code interact?

Metageographies of terrorism

We have already discussed the metageography of nation-states in Chapter 5. The meta-geography of a network contains two important components, nodes and conduits. The political outcomes of the network are a product of the actions of the people located at different nodes and the way they facilitate flows between nodes. For example, for a terrorist network to function money, people, weapons, explosives and other equipment, and information must move from node to node. The different nodes in a network will have different functions: training, gathering information, planning, finance, and execution of terrorist acts, for example. Terrorist networks are organized to minimize the amount of contact between nodes so that if one node is identified and engaged by counter-terrorist forces the whole network is not disrupted (Flint, 2003a). Terrorist groups have developed networks in this way over a number of years. For example, the IRA operated different cells of bombers on the British mainland without them knowing of each others existence. Al-Qaeda, as we shall discuss in greater detail later, is a different model, a network of loosely affiliated movements, perhaps best thought of as an "idea" or common cause than as an "organization" with its implications of central-ized control and bureaucratic hierarchy.

An abstract model of a terrorist network requires the definition of particular nodes (commonly referred to as "cells") and the connections (or flows) between them. A terrorist attack requires successful cooperation between cells located across the globe. What are the types of cells in a terrorist network? In what types of places are different types of cells located? How are the cells connected? These questions require the com-bination of the architecture of networks and the geography of places.

First, the structure of the network must be understood. What may be called "core nodes" are the cells that provide the highest level of planning and purpose of the network. "Peripheral nodes" are the cells that undertake the attacks, the bombers, hijackers, kidnappers, etc. In between are "junction nodes" that translate the plans into action by coordinating funding, training, recruitment, and equipping of the "peripheral nodes." Identifying and destroying the "junction nodes" will maximize the disruption of the network (Hoffman, 2002) because they are the most connected of all the nodes.

To target "junction nodes" they must first be located. The intersection of networks and territory determines particular categories of places that are most suitable for the different types of nodes. Core nodes may be located in territories where state authority is either weak or sympathetic to the terrorists' ideology: in the case of al-Qaeda southern Afghanistan and northern Pakistan, for example. On the other hand, peripheral nodes must exist and operate in relatively exposed spaces, or those where security is high: airports, borders, secure government and public buildings, etc. The nature of peripheral nodes and the environment in which they must operate makes their "appearance" brief. Also, though destroying a peripheral node will prevent a terrorist attack, the impact on the whole network is limited.

The junction nodes are not only the most connected in the network but also the most exposed; they must have a degree of permanency in relatively exposed spaces. Junction nodes must coordinate the logistics of the network, contacting forgers, arms sellers, smugglers, financiers, etc. To maintain such contacts requires a relatively stable presence in border zones and cities where security forces may be able to establish surveillance and enforcement presence. In other words, they are the most vulnerable and most important nodes in the network.

Terrorism expert Bruce Hoffman's (2002) identification of a hierarchy of al-Qaeda operatives does not explicitly address the geography of the network, but does point to the differential role of particular nodes. Hoffman identifies four levels of "operational styles." First is the professional cadre: the well-funded and "most dedicated, committed and professional element" of the group who are tasked with the most important missions. Second are the "trained amateurs" who may well be recruited from other terrorist organizations and have received some training. Their funding is limited and they are charged with "open-ended" missions, i.e. target US commercial aviation rather than a specific target. Third, are the local walk-ins: locally based Islamic radical groups who seek al-Qaeda sponsorship for their own projects. Fourth are the "like-minded insurgents, guerrillas and terrorists": the beneficiaries of bin Laden's financial "revolutionary philanthropy" and spiritual guidance. The relationship between these groups and al-Qaeda is mutual as they may also offer local logistical support for al-Qaeda operations.

Hoffman's hierarchy of al-Qaeda operatives provides clues into the spatial organization of a terrorist network. Key operatives are trained at particular nodes, and have access to money generated and distributed through another set of nodes. The "trained amateurs" have access to some training nodes but are denied the support of other nodes, especially finance, and so display less connectivity than the "professional cadre." Local logistical support can also be "outsourced" to the "like-minded," preventing the need for all support to come from what could be termed an al-Qaeda network.

What are the implications of such a network organization for counter-terrorism? Hoffman's recommendations reflect an implicit recognition of a hierarchy of nodes in a network. The first recommendation is to target "mid-level leaders" as "[p]olicies aimed at removing these mid-level leaders more effectively disrupt control, communications, and operations up and down the chain of command" (Hoffman, 2002, p. 21). In other words, these leaders staff important nodes in the network, which facilitate the combination of plans and resources that make a terrorist attack happen. In network terms, Hoffman is proposing the targeting of a junction node that once gone negates the efficacy of all other nodes.

Hoffman's second recommendation is to "[d]e-legitimize—do not just arrest or kill—the top leaders of terrorist groups" (Hoffman, 2002, p. 22). The argument being that leaders do more than coordinate a network they give ideological purpose to its existence. By portraying the leader as corrupt or hypocritical the ideological glue binding the network together may loosen.

The third recommendation is to "[f]ocus on disrupting support networks and trafficking activities" (Hoffman, 2002, p. 22) The terrorist requires a network of support, if these supporting connections are disrupted (and they may be easier to identify and arrest) then the final node of the network is starved of what it needs. Hoffman's fourth recommendation is to "[e]stablish a dedicated counter-intelligence center specifically to engage terrorist reconnaissance" (Hoffman, 2002, p. 23). Reconnaissance may either be the sole task of a particular node or one of the tasks of the ultimate perpetrators, but it requires a degree of visibility at what is likely to be a well-policed location. These last two counter-terrorism recommendations recognize that certain nodes are more vulnerable than others, and make for more profitable counter-terrorism.

It is not just the function or type of the node that is crucial, it is also the geographic context in which it operates. For example, Hoffman's (2002) recognition of reconnaissance activities is given further import because of the need for a terrorist to spend time in a well-policed location. The coordinating role of mid-level leaders may require a certain fixity and visibility at a particular location that abets counter-terrorism. On the flip side, the ideological function of leaders allows them to retreat to geographical areas that are hard to police: the tribal areas of Pakistan, for example. Finally, the merging of terrorist networks with other criminal activities, such as smuggling, requires terrorist networks to operate in border zones that may facilitate counter-terrorism. The geography

Figure 7.3 War on Terror.

of the terrorist network is laid over maps of policed territories, and the variation in the level of policing across space. Terrorists try to locate nodes with this geography of policing in mind. Counter-terrorist agencies try to identify where nodes are forced to become the most visible (Figure 7.3).

Terrorists have created a metageography of the terrorist network in order to fight power organized in a different and established metageography, territorial sovereign states (Flint, 2003a and b). In the first three waves of terrorism, networks were mainly organized within a particular state, hence the jurisdiction of counter-terrorist forces overlay the spatial extent of the network. However, during the third wave of terrorism this geographical relationship began to change as training, especially, was conducted in foreign countries. Cooperation between states (such as that between France and Spain to counter ETA) was relatively easy as they were neighbors with a common interest against the terrorist group. The internationalization of the PLO was a different matter, operating in either states or territories that did not facilitate cooperation between states. The current War on Terrorism has made the situation much harder for states. The goals of al-Qaeda are hard to discern and the geography of the network has been difficult to identify. Even when it appeared that operating cells within the US were identified, some of these allegations have not stood up to judicial scrutiny.

Case study 7.1: al-Qaeda, its history and metageography

This case study will provide background information on the growth and activities of al-Qaeda before presenting a discussion of its geographical expression. The aim is to illustrate that al-Qaeda is not the centralized and singular entity that it is often portrayed to be by governments fighting the War on Terrorism. Indeed, the metageography of al-Qaeda seems to be aligned to sub-national territorial struggles, with a global *jihad* against the US being a rather loose ideological frame. The timeline is constructed from a number of sources listed at the end of the book. We will begin the story in a turbulent Afghanistan during the Cold War.

Timeline

1973   July 17—The Soviet-backed Mohammed Daoud overthrows King Zahir Shah and proclaims Afghanistan a republic.

1978   April 27—The People's Democratic Party of Afghanistan kills Daoud in a coup.

- The new president, Noor Mohammed Taraki, makes attempts to impose land reform and mandatory education for women; these new policies "spark a nationwide *jihad*" (Griffin, 2003, p. xvii).

1979   The Soviets invade Afghanistan to support Taraki.

- Osama bin Laden enters Afghanistan to help the effort of the *mujahideen* fighting the Soviets.

- Osama bin Laden was born in Saudi Arabia in 1957 to a Yemeni father (to whom he was the seventeenth child of 51) and a Saudi mother. His father, Mohammed bin Laden, was founder of a construction company, the Bin Laden Group, which made the family a billion-dollar fortune.

- During the time of the Soviet invasion, Osama bin Laden was studying at university in Jeddah, Saudi Arabia, where he earned a degree in civil engineering and public management.

1988   February—Gorbachev announces a ten-month phased withdrawal of Soviet troops to begin in May.

■ This same year al-Qaeda is founded by bin Laden, Mohammed Atef, and Abu Ubaidah al Banshiri—the headquarters are located in Afghanistan and Peshawar, Pakistan (Alexander and Swetnam, 2001, p. 37).

1989   After the Soviets have evacuated Afghanistan, bin Laden moves to Saudi Arabia.

1991   When American troops move into Saudi following the Iraqi invasion of Kuwait, bin Laden moves to Khartoum, Sudan.

■ "Distraught by the American presence near the holy cities of Mecca and Medina, Bin Laden decides to move his base elsewhere" (Alexander and Swetnam, 2001, p. 38).

■ "Al Qaida's global network, as we know it today, was created while it was based in Khartoum, from December 1991 till May 1996" (Gunaratna, 2002, p. 95).

- "Al Qaida enjoyed the patronage of the Sudanese state until the . . . mid-1990s" (Gunaratna, 2002, p. 158).

1992   December 29—The first official al-Qaeda attack takes place at a hotel in Aden, Yemen.

■ "Intended to kill US troops en route to Somalia on a U.N. relief mission, but the troops had already left the premises" (Alexander and Swetnam, 2001, p. 39).

1993   February 23—Truck bomb at the World Trade Center in New York kills 6 and injures over a 1,000.

■ In November of 1997 Ramzi Yousef and Eyad Ismoil, two al Qaeda members, are found guilty of the attack.

October 3–4—Two Black Hawks are shot down in Mogadishu.

■ "US subsequently learns that Bin Laden's organization had been heavily engaged in assisting warlords who attacked US forces in Somalia" (BBC, 2004a).

1994   October—Kandahar falls to Taliban, led by Mullah Mohammed Omar.

■ Two al-Qaeda operatives, Wadih el Hage and Mohammed Sadeek Odeh, move to Nairobi and Mombasa, Kenya respectively—they open multiple businesses with al-Qaeda funds (Alexander and Swetnam, 2001, p. 40).

1995   June 26—Attempted assassination of Hosni Mubarak, the Egyptian president, in Addis Ababa.

- Mubarak was cracking down on extremist Muslim groups in Egypt, and the attack is therefore thought to be al-Qaeda related (Alexander and Swetnam, 2001, p. 33).

November 13—A car bomb kills five Americans and two Indians in Riyadh, outside of an American-operated Saudi National Guard training center.

1996   June 25—A truck bomb outside of a US Air Force housing complex in Dhahran, Saudi Arabia kills 19 US soldiers. Al-Qaeda involvement is suspected.

September 27—Taliban seize Kabul.

- Al-Qaeda moves from Khartoum back to Afghanistan where the Taliban offer a safe haven (Alexander and Swetnam, 2001, p. 40).

1998   Bin Laden and various fundamentalist groups from Pakistan and Bangladesh endorse a *fatwah* that "states that Muslims should kill Americans, including civilians, wherever they can be found" (Alexander and Swetnam, 2001, p. 41). This is followed by numerous other similar declarations.

August 7—Bombings of US embassies in Kenya and Tanzania.

August 20—US retaliates with several missile attacks on bases in Afghanistan and the bombing of a suspected al-Qaeda chemical plant in Sudan.

1999   Al-Qaeda base of operation is moved from Kandahar to Farmihadda, Afghanistan (Alexander and Swetnam, 2001, p. 44).

July 30—Taliban "reinforce its position to offer Osama bin Laden safe haven ..." (Alexander and Swetnam, 2001, p. 45).

October 16—UN "imposes sanctions on the Taliban until Bin Laden is expelled from their borders" (Alexander and Swetnam, 2001, p. 45).

2000   June—Pakistan makes claims that the Taliban have shut down al-Qaeda training camps (Alexander and Swetnam, 2001, p. 48).

October 5—17 American sailors are killed by suicide bombers aboard the USS Cole in Yemen.

2001   September 11 attacks on the World Trade Center and Pentagon.

2002   November 28—Attacks in Mombasa, Kenya.

- Two different attacks targeted Israeli tourists:
  - A missile just missed a Boeing 757 heading to Tel Aviv.
  - A car bomb exploded outside of a hotel taking the lives of 13 victims plus the 3 bombers (CNN, 2002).

2003   May 12—Four near simultaneous suicide bombings occur at compounds that house Westerners in Riyadh.

- 23 victims and 12 bombers killed.

- "Ali Abd al-Rahman al-Faqasi al-Ghamdi, who authorities said has deep ties to al Qaeda, surrendered" in connection with the planning of these bombings (CNN, 2003).

2004   March 11—Al-Qaeda claim responsibility for coordinated bombing attacks on commuter trains in Madrid, Spain, citing the deployment of Spanish troops in Iraq. Some 191 people were killed and 1,460 injured.

2005   July 7—London, Great Britain. In the morning rush hour three bombs are detonated by suicide bombers simultaneously in separate locations on the London Underground. An hour later, a fourth suicide-bomb is detonated on a bus. Some 56 people, including the bombers, are killed and 700 injured. This is the first incident of suicide bombing in Western Europe. Bombers are believed to have had ties to British Islamist groups connected to al-Qaeda.

July 21—London, Great Britain. Four attempted bombings in the London Underground system, but only the detonators explode. No fatalities or injuries. Four suspects subsequently arrested.

The metageography of al-Qaeda

Following other terrorist organizations, al-Qaeda is made up of separate cells that can vary greatly in size. Some cells consist of only two people in a location and the network is designed so that the arrest or destruction of one cell would not affect the others: "Cells assigned for special missions like 9/11 . . . are coordinated through an agent-handling system where a cell leader reports only to his controller or agent-handler. Most agent-handlers live near the target location or in the 'hostile zone'—Europe or North America" (Gunaratna, 2002, p. 97). In other words, the controller provides the connection between the cells, or nodes of the network so that knowledge of the connectivity of the network is limited.

For example, the cells used as "launchpads" for the 9/11 attacks were independent ones operating in the United Arab Emirates, Germany, and Malaysia: "They were secured by strict compartmentalization but a few select members were permitted to liaise between the compartmentalized cells" (Gunaratna, 2002, p. 104). Furthermore, the network was not determined by strict centralized control. "Individual Groups or cells appear to have a high degree of autonomy, raising their own money, often through petty crime, and making contact with other groups only when necessary" (BBC, 2004b).

There was an identity or organization to the network based upon a different form of metageography, national identity. Members of al-Qaeda operated within the "family" of their national group: with the families divided by function as well as by place of origin. For example, "Al Qaida's Libyans managed the documentation and passports office in Afghanistan; its Algerians ran fraudulent credit card operations in Europe; and the Egyptians looked after most of the training facilities worldwide" (Gunaratna, 2002, p. 97).

The al-Qaeda network overlays other political geographies in another way too. Rather than being a network operating under one determining ideology or mission, the network cells are established through ties with other like-minded groups that are already in existence in various parts of the world. The ideological views shared by al-Qaeda and smaller localized terrorist organizations are mainly structured around Islamic law. In terms of a geopolitical code, "Westerners" are seen as the enemy partly because of their actions and partly due to propaganda promoted by certain groups that influence young Muslims to join in the *jihad* against them.

Al-Qaeda's proclaimed objectives in summary are:

- Oppose all nations not governed according to their particular interpretation of Islam

    - i.e.) US—secular law
    - i.e.) Saudi Arabia—misinterpreted Islamic law, according to Al Qaeda

- Oppose the presence of US forces in the Middle East, especially near Mecca and Medina
- "Attack the enemies of God"
- Eventually create a unified Muslim nation-state (stretching from Spain to South East Asia)

(Alexander and Swetnam, 2002, p. 2)

Establishing ties with already-formed groups aids al-Qaeda in its avoidance of counter-terrorist organizations whose jurisdiction is based upon the metageography of state boundaries. Cells based in the local culture and landscape help al-Qaeda to recruit and operate in places where central government actions are hostile, but do not have popular support, for example northern Pakistan. Furthermore, in the wake of the increased policing of boundaries "Osama and al Zawahiri realized that establishing a network from scratch would not be easy and therefore decided upon ideological infiltration" (Gunaratna, 2002, p. 114). In other words, the cells of the network and form of connectivity were associated with the "opportunities" provided by local conflicts and capabilities.

Recent examination of Southeast Asia terrorist groups indicates, for example, that they have had ties with al-Qaeda for many years.

Authorities are now unearthing and piecing together evidence that, far from being locally-contained separatist groups, many terrorist organizations in the region in fact have close and long-running connections not only with each other, but to Osama bin Laden's *al Qaeda* as well.

(Huang, 2004)

These Southeast Asian groups are thought to have been bin Laden's portal to the area, helping to expand the influence of al-Qaeda and to help the group set up cells of its own: "Both Jemaah Islamiah and KMM [Kumpulan Militan/Mujahideen Malaysia] have cells throughout Southeast Asia, and their Afghan-trained members are believed to have served major roles in expanding the *al Qaeda* network in the region" (Huang, 2004).

The metageography of the al-Qaeda network is a combination of separate cells, many formed with regard to territorial conflicts, but held together by an overarching ideology. Indeed, it is necessary that groups share the same beliefs and goals of al-Qaeda in order for a common ideology to cement the network. The Southeast Asian group Jemaah Islamiah's spiritual leader, Bashir, tells his followers that the west and Zionism "have been plotting for decades to destroy Islam and to dominate the world" (Huang, 2004). There is also evidence that al-Qaeda has links with terrorist groups in Kashmir, Uzbekistan, Philippines, Algeria, among others (BBC, 2004b).

The metageography of al-Qaeda has been forced to adapt in response to the geopolitical codes of the US, and other countries, after the terrorist attacks of September 11, 2001. The US invasion of Afghanistan and the fall of the Taliban forced al-Qaeda to construct its network metageography in an even looser form. It is now "structured in a way that it can react very quickly to changing events on the ground. Mobility, flexibility and fluidity will be the guiding principles" (Gunaratna, 2002, p. 96). Perhaps the term "al-Qaeda network" is increasingly a geopolitical fiction: a handy representation to justify a global War on Terrorism when, in fact, the cells of the network are looser than ever. Indeed, "[s]ome analysts have suggested that the word al-Qaeda is now used to refer to a variety of groups connected by little more than shared aims, ideals and methods" (BBC, 2004b).

It is useful to consider al-Qaeda's metageography as a network of territorially based and separate cells. The "local" basis of these cells provides strength for the network in terms of its ability to operate effectively across the globe, beyond the reach of the state security apparatus. However, the combination of territorial and network metageographies illustrates potential problems for al-Qaeda: is bin Laden's geopolitical code coherent and strong enough to dominate diverse local concerns, and how susceptible are the local groups to negotiations with their respective states that would undermine their commitment to a global project?

Incongruous geographies?

The larger metageographic point is that in order to counter a terrorist network, the United States has had to conquer sovereign territory (Flint, 2003b). The methods of terrorism and counter-terrorism construct very different, even incongruous, geographies that have implications for the success of counter-terrorism. States must challenge networks by controlling sovereign territory. More than just being inefficient, this may actually be a counter-productive counter-terrorism as it increases the presence of US forces in other countries. As a result, bin Laden's *fatwa* becomes prophetic.

The primary purpose of the invasion of Afghanistan and the overthrow of the Taliban regime was the disruption of al-Qaeda bases: a sovereign state was invaded to destroy the nodes of a network. The operation has been only partially successful as some territory has remained beyond the control of the US and allied security forces, namely eastern parts of Afghanistan and northern parts of Pakistan. Moreover, the US's territorially based response to the attacks of September 11, 2001 have reinforced the rhetoric of al-Qaeda that views the United States as conducting a global "crusade" against Muslims. The strategy of controlling territory to combat a network has not only reinforced the perceptions of al-Qaeda sympathizers that the US is on a global mission, but has also relocated US troops and made them potential targets (refer back to al-Qaeda's geo-political code on p. 70). Figure 7.4 shows the extension of US bases into central Asia immediately after the terrorist attacks of September 11, 2001.

The same strategy was used to justify the 2003 invasion of Iraq, though subsequently President George Bush's administration admitted there were no connections between al-Qaeda and Saddam Hussein. Saddam Hussein was portrayed as a ruler who was using the territorial sovereignty of Iraq to facilitate the maintenance of the al-Qaeda network. Justification for the war rested upon the need to invade the sovereign territory of Iraq to disrupt a network that had some, poorly defined, connection with Iraq, and may use those connections to conduct further attacks within the sovereign territory of the United States. Simply put, the US argued that disrupting a terrorist network required the mili-tary invasion and occupation of sovereign territory.

However, it is also evident that the War on Terrorism is using less territorial tactics to counter the terrorist network of al-Qaeda. First, cooperation with other countries has met with some success as arrests of alleged terrorists have been made in Pakistan and Indonesia for example (see Box 7.5). Less conventional, and with greater geopolitical implications, is the use of aterritorial weaponry to target alleged terrorists in other sover-eign spaces. In November 2002 an unarmed US drone fired a missile at a truck in the Yemeni countryside, killing six people identified as terrorists linked to al-Qaeda. The ability of the United States to act within the sovereign spaces of other countries is related to its position as world leader. In the next section we interpret the geography of the War on Terrorism within the cycle of world leadership.

Box 7.5 Sovereignty and counter-terrorism

On August 14, 2004 the *Economist* reported that information used to raise the US terrorist threat level had come from computer files found in Lahore and Gujarat in Pakistan after the arrest of two men, one a Tanzanian. The raid led to the arrest of 13 Muslim men in Britain. It is evident that counter-terrorism is a matter of coordination between national law enforcement agencies. Such international coordination relates to the identification of "most-wanted terrorists" by the US. A selection of eight arrests reported by the *Economist* reflects the relationship between terrorist networks and the jurisdictions of sovereign states. Listing first the suspects nationality and then the country in which they were arrested: Pakistani—Pakistan; British—Britain; Saud—Pakistan; Yemeni—Pakistan; German—Morocco; Saudi—Saudi Arabia; Saudi—United Arab Emir-ates; Moroccan—Germany.

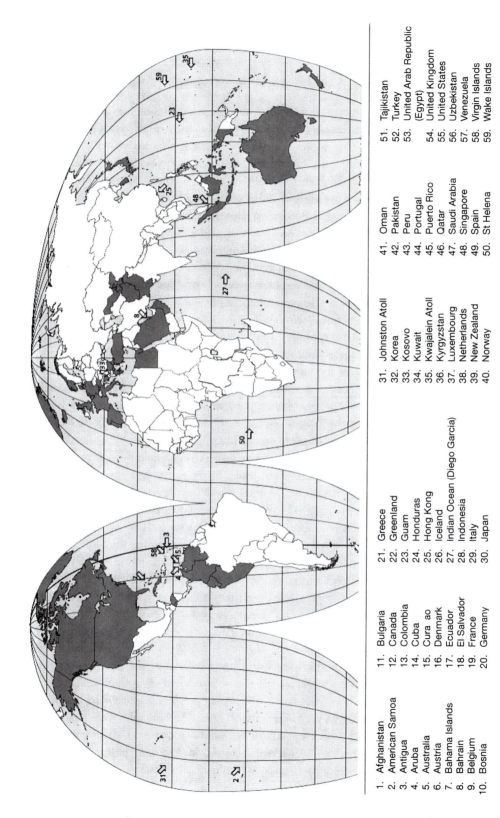

1. Afghanistan
2. American Samoa
3. Antigua
4. Aruba
5. Australia
6. Austria
7. Bahama Islands
8. Bahrain
9. Belgium
10. Bosnia

11. Bulgaria
12. Canada
13. Colombia
14. Cuba
15. Cura ao
16. Denmark
17. Ecuador
18. El Salvador
19. France
20. Germany

21. Greece
22. Greenland
23. Guam
24. Honduras
25. Hong Kong
26. Iceland
27. Indian Ocean (Diego Garcia)
28. Indonesia
29. Italy
30. Japan

31. Johnston Atoll
32. Korea
33. Kosovo
34. Kuwait
35. Kwajalein Atoll
36. Kyrgyzstan
37. Luxembourg
38. Netherlands
39. New Zealand
40. Norway

41. Oman
42. Pakistan
43. Peru
44. Portugal
45. Puerto Rico
46. Qatar
47. Saudi Arabia
48. Singapore
49. Spain
50. St Helena

51. Tajikistan
52. Turkey
53. United Arab Republic
    (Egypt)
54. United Kingdom
55. United States
56. Uzbekistan
57. Venezuela
58. Virgin Islands
59. Wake Islands

Figure 7.4 Geography of US bases.

World leadership and the War on Terrorism

Rewriting the geographic understanding of global politics?

It is relevant here to turn back to Chapter 2 and our discussion of world leadership. As world leader, the United States is the main target of contemporary terrorism and organizer of a coalition of states in the War on Terrorism. It also has the military power, especially in terms of technological capacity, to conduct such extra-territorial actions. These range from the establishment of military bases in other countries, flying, using Special Forces to train and coordinate armies across the globe, cooperating with foreign police forces, as well invading countries, see Box 7.2 (Priest, 2003). On the other hand, its presence across the globe (in terms of business, diplomacy, and the military) provides numerous targets. Terrorists opposed to the United States' extra-territorial power can also find targets because of that extra-territoriality.

The power of the world leader rests not in its military strength alone, but rather on the package of innovations it claims will benefit the whole world. The central ingredients of this package are national self-determination and democracy, or the rule of law. Together these "innovations" combine to form the integrative power of the world leader: the collection of ideas, values, and institutions designed to bring order and stability to the world. The targeting of the United States by al-Qaeda is a violent challenge to its world leadership role, as we discussed earlier. More important is the fundamental underlying issue that while facing that challenge by defining and conducting the War on Terrorism, the United States is violating the very ideals it claims are the basis of the innovations that are the foundations of its world leadership.

The November 2002 execution of terrorist suspects in Yemen by an unmanned US drone exemplifies the ability of the world leader to project its power across the globe: in the form of troops stationed in the country as well as the technological capacity to see and strike beyond the horizon. The actual amount of awareness the Yemeni government had of this attack is not clear; they had good reason to deny any knowledge to placate domestic groups hostile to any government cooperation with the US. Conducting summary executions in other sovereign spaces signifies that the world leader is reinterpreting the territorial understandings that are the foundation of the political geography underpinning its world leadership, state sovereignty and the rule law. The legitimacy of the world leader, or the strength of its integrative power, is challenged if it is seen to be breaking the very rules and values it argues for.

The threat to the world leader in the practice of the War on Terrorism is not just a matter of its integrative power. If the world leader condones and practices an extra-territorial geopolitics then these tactics may diffuse to other countries. Russia and Israel have voiced their belief that they may conduct preemptive attacks on unilaterally defined terrorist targets given the United States' lead. The principle of the violation of other countries' sovereign space to conduct preventive, preemptive, or revenge attacks can legitimate state upon state violence across the world. The importance of sovereignty in international law rests upon the gravity of violation of a country's borders: the goal being to prevent violence, and the escalation to war. However, the killing of another country's citizens in their own territory as a matter of summary justice violates the assumptions of international law and undermines any judicial consideration of who is guilty and how they should be punished. The role of sovereignty in protecting citizens and putting the brakes on international violence is undermined.

The potential for increased conflict around the world is the opposite of the world leader's goals of order and stability. If increased conflict was to occur, the military demands of the United States across the globe are likely to increase. As we discussed in Chapter 2, such "imperial overstretch" weakens the coercive and integrative power capabilities of the world leader, resulting in increased challenge and decline.

There is another way to look at the War on Terrorism, though. In the words of the Bush administration in their response to 9/11, the world has changed and requires a redefinition of war. The world leader is redefining our understanding of territorial sovereignty in the face of the new challenge of global terrorism to global order. In this interpretation, the US is playing the necessary and useful role of a world leader, making political and geographic innovations to provide stability. In this interpretation, the War on Terrorism is seen by US policy-makers as a response to violent challenges of its authority, but also serves as an "opportunity" (in the words of Defense Secretary Rumsfeld) to advance the world leader's agenda.

However, this interpretation may be looked at in a different way too. At a time of violent challenge to its world leadership role, the United States has had to legitimize its reaction. The reaction violates commonly held views of how global politics is organized geographically, and will provoke greater challenge to the world leader and further undermine its power. For example, UN Secretary-General Koffi Anan has been scathing, well as scathing as a UN diplomat can be, toward what he sees as a cavalier approach to the geographic axioms of world politics. To return to the language of Modelski's model (1987), responding to violent challenges (deconcentration) has also intensified the diplomatic challenges (or delegitimation) facing the United States.

There is yet another interpretation, that the War on Terrorism is a created threat— exaggerated to serve the purposes of states, especially, but not exclusively, the United States. In this argument, the capabilities of al-Qaeda are seen to be over-exaggerated in order to justify military action overseas and also, such as in the case of Chechnya (Chapter 5), repression of groups within the state. This interpretation follows the work of Giorgio Agamben (1995) and his discussion of the Roman designation of *homo sacer*—a category of people deemed beyond the sacred (and so not worthy of sacrifice to the gods) but also outside juridical law (and so could be killed with impunity). The mixture of profane and beyond legal protection results in categories of people that may be killed with little thought, guilt, or control by established institutions. Gregory (2004) argues that the representation of the War on Terrorism as a crusade of good against evil has facilitated geopolitical spaces where the US can use its overwhelming firepower with impunity, notably in the invasions of Afghanistan and Iraq. The War on Terrorism is viewed as a construct that facilitates the global presence and military actions of the world leader, and also justifies killing large numbers of people, many—if not most— civilians, in the name of "global order."

Global visions technologically enhanced

The second major implication of the way the War on Terrorism is being fought lies in the usage of surveillance strategies, and their connection to remotely controlled weaponry. As we discussed in Chapter 1, a key component of modern geopolitics is the ability to envision the whole globe. The desire of geopoliticians has always been to "map" the world; mapping in this sense is a proclamation of having understood the basic processes and features of the world. In other words, a claim to knowing how the world works. From the earliest cartographic visions to the creation of "theories" of world conflict and the "natural" behavior of states, geopoliticians have always proclaimed an ability to "see" the world as a unified entity with identifiable structures and processes. The trick of the geopolitician was to portray these understandings as being "objective" and then using them to justify or exhort particular actions for their state: Mackinder's imperial urges and Mahan's calculated need for a strong navy, for example.

The technological advances in surveillance have changed the imperative to "map" through theory. The world can be observed to a much greater degree now than in the past. Real-time observations across the globe are possible through satellite images. The global vision is technologically possible. What is still required is the ideological under-standings that can make sense of the map, and it is here that the representation of geopolitical codes remains a vital component of the War on Terrorism.

Summary and segue

The interrelated geographies of the fourth wave of terrorism and the War on Terrorism have illuminated some key points. The geography of contemporary terrorism manifests itself in both interstate cooperation and ideology that transcends the state while still retaining connections to the state as a geopolitical structure. The internationalization of terrorism requires a consideration of the interaction between two metageographies: networks and states. It is not the case of network geographies usurping or eradicating state geopolitics, but an antagonistic relationship between the two. Terrorist networks require, to some degree, the territorial sovereignty of states and state counter-terrorism cooperates across boundaries and also may violate them. Finally, the War on Terrorism is a geopolitics of these metageographies that may be interpreted within the broader structure of world leadership.

Identifying the geography of terrorism and counter-terrorism illuminates that there are many uncertainties and unknowns within a structural approach to geopolitics. The form of agency is far from determined. All that a structural approach can do is suggest some parameters for what may occur and an overarching context to try to explain what does happen. The concluding chapter explores the "messiness" of geopolitics further.

Having read this chapter you will be able to:

- consider the geopolitics of globalization;
- locate the geopolitics of globalization within Modelski's model;
- identify the geography of contemporary terrorism;
- identify the geography of contemporary counter-terrorism;
- consider the geographic mismatch of terrorism and counter-terrorism;
- situate anti-US terrorism and the War on Terrorism within Modelski's model of global geopolitics.

Further reading

Flint, C. (2005) "Dynamic Metageographies of Terrorism: The Spatial Challenges of Religious Terrorism and the 'War on Terrorism'" in Flint, C. (ed.) *The Geography of War and Peace*, Oxford: Oxford University Press, pp. 198–216.

A discussion of terrorism and the War on Terrorism emphasizing the interaction between network and nation-state metageographies.

Gregory, D. (2004) *The Colonial Present*, Oxford: Blackwell Publishing.

A passionate and provocative analysis of contemporary conflicts emphasizing the way states, especially the United States and Israel, commit violence upon civilians.

Hoffman, B. (1998) *Inside Terrorism*, New York: Columbia University Press.

An excellent and accessible introduction to the study of terrorism.

Juergensmeyer, M. (2000) *Terror in the Mind of God*, Berkeley, CA: University of California Press.

A thought-provoking analysis of the motivations and implications of terrorism conducted by religious fundamentalists in all the major religions.

Priest, D. (2003) *The Mission: Waging War and Keeping Peace with America's Military*, New York: W.W. Norton and Company.

A *Washington Post* journalist travels the globe with the US army showing its many roles in foreign countries, an eye-witness account of extra-territoriality.

MESSY GEOPOLITICS: AGENCY AND MULTIPLE STRUCTURES

In this chapter we will:

- emphasize the complexity or "messiness" of geopolitics. In other words . . .
- highlight the interaction of multiple geopolitical structures in creating specific geopolitical contexts;
- focus on the topic of rape as a weapon of war to illustrate the argument;
- use a case study of the conflict over Jammu and Kashmir to exemplify the concepts;
- note how complexity or "messiness" is a product of the interaction of structure and agency;
- situate "messiness" within the structure provided by Modelski's model.

This chapter will conclude our introduction to geopolitics by emphasizing complexity, or "messiness." Each of the previous chapters has focused upon a particular set of geopolitical agents and structures: the geopolitical codes of states, or the metageography of terrorist networks for example. However, in Chapter 1 we introduced agents and structures by talking about how they could be seen as nested scales; in other words any geopolitical agent will have to simultaneously negotiate the opportunities and constraints of a number of structures. Furthermore, geopolitical agents have multiple goals—they are not homogenous, simple, or singular entities. The multiplicity of geopolitical goals is evident in individuals, nations, states, terrorist groups, and any other geopolitical agent. In other words, geopolitical agents juggle a number of identities, some competing and some complementary. Combining the multiplicity of agents' identities and goals with the combination of geopolitical structures indicates that geopolitics is a messy affair.

Who am I, who am I fighting, and why?

The cause of the Palestinians is commonly identified as a nationalist struggle—the desire of a people for their own independent state. Indeed, this was the focus of the case study in Chapter 5. However, to talk of the Palestinians as a homogenous group is false. On the one hand, the politics of the national group is the product of competing groups with different goals. For one, factional politics is a key part of Palestinian politics. But instead of focusing on formal or party politics, let us begin with the example of an individual Palestinian man, 'Adnan, living in the Rafah refugee camp in June 2001, and his experiences of an Israeli army raid that demolished 17 houses, the homes of 117 people:

> While the shelling continued, I took my disabled mother, who requires a wheelchair, and told my wife and children to get out of the house. They were all frightened and hysterical. Throughout the neighborhood there were screams of little children, and adults asking, "Where is my son? Where is my brother? Did they get out?" [. . .] At approximately 5:30 A.M. it ended. The army left the area, and I looked for my wife and children. My sister Hanan told me that my wife, who is pregnant, was on the main road and couldn't stand on her feet out of fear because of the horrible sight of the demolished houses. I went to her and asked what happened. She said that she was bleeding, a result of fear and the running from the house. [. . .] The army also demolished my irrigation pool, the shed with motors and pumps, and a one-hundred-square-meter sheep pen. The pen had six sheep and one of them was killed during the demolition. The bulldozer also uprooted six olive trees that were forty years old.
>
> (B'TSELEM, 2002, 17–18; quoted in Falah and Flint, 2004, p. 124)

Who is 'Adnan, or in what ways can we identify him as a geopolitical agent? Father, father-to-be, husband, son, brother, farmer, and current guardian of an olive grove that would be the hope of income for future generations. At the intense moment of the destruction of his home, what are 'Adnan's geopolitical goals? The quote stressed protection of his immediate family, both in the sense of their physical health, as well as their economic well-being. Family and economics are the structural imperatives in the quote. Of course, they are linked to his plight as a refugee, and so to his membership of a stateless nation. The limited efficacy of his agency must be understood in relation to the coercive power of the Israeli Defense Forces, as well as the metageography of global geopolitics; legitimacy is accorded to states and their citizens have rights that are not possessed by refugees.

The next example is also intended to illustrate that geopolitical agents operate within a number of geopolitical structures, even if they are not conscious of them (Figure 8.1). Darfur is a region of western Sudan, and home to ethnic Africans. The region has been in conflict with the Arab dominated Sudanese government. Calls have been made to classify the killings in the region as genocide sponsored by the Sudanese government. The international community has been reluctant to label the killings genocide, as widespread and systematic as they are, because it would produce expectations of intervention. The violence has been committed by Arab militias known by their victims as *janjaweed*, or "devil's on horseback," widely believed to be doing the bidding of the government, though this is denied by officials in Khartoum. The UN investigation of the attacks has highlighted the use of rape as a weapon by the *janjaweed*. The reported incidents are numerous, but to give a sense of the horror one attack in March 2004 involved the rape of 16 girls by 150 soldiers and *janjaweed*. It is alleged that girls as young as 10 years old have been raped.

The geopolitical implications of rape as a weapon of war are discussed in Box 8.1. Here the topic of multiple geopolitical structures is illustrated by a February 11, 2005 article from the *New York Times* that reported on the difficult future of the babies that are being born as an outcome of systematic rape in Darfur. Fatouma, a 16-year-old mother and rape victim, identifies her baby as a *janjaweed*. "When people see her light skin and her soft hair, they will know she is a *janjaweed*" (Polgreen, 2005, p. 191). For now, the baby is being raised and protected, but the future for both mother and baby is uncertain given the deep cultural taboos regarding rape and, in the Muslim tradition, identity is passed from father to son. The article continues by quoting the village sheik's thoughts toward the baby:

> She will stay with us for now . . . We will treat her like our own. But we will watch carefully when she grows up, to see if she becomes like a *janjaweed*. If she behaves like a *janjaweed*, she cannot stay among us.

Ethnic identities of them and us, as well as the position of women in a patriarchal and traditional society interplay to make the future for women such as Fatouma and her

Figure 8.1 Child soldiers.

Box 8.1 Rape as a weapon of war

Rape is increasingly used as weapon of war: It "routinely serves as a strategic function in war and acts as an integral tool for achieving particular military objectives" (Ramet, 1999, p. 206). Recent and ongoing conflicts in the former Yugoslavia, Rwanda, Darfur/Sudan, Burma/Myanmar, Jammu and Kashmir have all involved systematic rape. Rape is an effective weapon because it has an impact upon a number of geopolitical structures and, hence, is disruptive in many ways.

For example, Allen's (1996) discussion of rape in the former Yugoslavia points to the ability of systematic rape to change the perception of place; after public rapes of Bosnians and Croats had demonstrated the danger of remaining, people would leave their established homes and leave the vacated town for occupation by Serbs. Rape had changed a place from a traditional site of community to a venue of fear, and so facilitated the brutal redrawing of the ethnic geography of former Yugoslavia.

Rape in warfare is also a means of enforcing pregnancy "and thus poisoning the womb of the enemy" (Crossette, 1998). From this perspective the target of the rapists is at the individual scale of the mother and the offspring. The woman becomes "damaged goods in a patriarchal system that defines woman as man's possession and virgin woman as his most valuable asset" (Allen, 1996, p. 96). As one Rwandan rape victim said: "We are not protected against anything . . . We become crazy. We aggravate people with our problems. We are the living dead" (Human Rights Watch, 1999, p. 73). Rape victims are unable to find husbands and bear other children, and hence become rejected by their families and communities. The target of the rapist in war is also the child in a context in which membership of one ethnic community is vital and children born from rape can be seen, for example, as infusing Serbian blood into other ethnic groups and producing "little Chetniks" or "Serb soldier-heroes" (Allen, 1996, p. 96). Rape destroys the life of the individual and disrupts the identity and cohesion of the community and the ethnic group.

Key to understanding the ability to motivate soldiers to rape as well as the disruption of communities is the notion of patriarchy. The ability to violate and harm women, to see rape as an acceptable form of combat, requires soldiers to be socialized within structures that see the domination and control of women as a norm. The strategic understanding that rape victims will be rejected by their communities and families also rests upon the patriarchal view of women as "property" that cannot be married off or produce wanted children after rape.

Understanding the power of patriarchy is crucial in making sense of the impact of systematic rape, and hence its adoption in war. In a nationalist or ethnic conflict, when it would appear that group identity is the dominant geopolitical factor, a daughter attacked by the enemy group does not receive the sympathy and help of their own community. Patriarchal values trump communal solidarity. In the former

Yugoslavia, for example, women feared that they would be shunned by family and friends (Allen, 1996, p. 70), and the victims' trauma was "exacerbated by cultural taboos associated with rape" (Human Rights Watch, 1999). In Jammu and Kashmir, Pandit and Muslim women who suffered rape in the conflict were taunted by their neighbors (of their own cultural group) and sometimes outcast by their families. After rapes committed by the Indian security forces in 1991, "women had been deserted by their husbands . . . a seventy year old woman had been thrown out by her son . . . [and] girls . . . were teased even by the village men" (Chhachhi, 2002, p. 200). National and community solidarity in the face of conflict took second place to embedded views of the status of women. However, the very rejection of women rape victims by their own communities disrupts societies and cultures, and so is seen as an effective weapon of war.

The final geopolitical structure I will introduce in this discussion is the state. The case of Myanmar/Burma is especially indicative, though certainly not the only case, in which the government is active in promoting its soldiers as rapists. Systematic rape by the army is based upon a patriarchal society, with "many indicators of male predominance and female subordination throughout Burmese society" (Apple, 1998, p. 26). Also, the army is alleged to "recruit" teenagers by kidnap, and one argument is that systematic rape by Burmese soldiers is indicative of the abuse they have suffered themselves (Bernstein and Kean, 1998, p. 3). The Burmese government uses the army in its attempt to dominate minority ethnic groups. Similar to other conflicts, rape in Burma is used to illustrate the power of the state over ethnic minorities, to instill fear, and nullify any plans for rebellion (Women's Organization from Burma, 2000, p. 27). Furthermore, Burmese soldiers are taught that by impregnating women from ethnic minorities they will be leaving Burmese blood in the villages, which will end the rebellion. Perhaps unique to the case of Burma, and indicative of the combined domination of state apparatus and patriarchy is the belief that rape provides the opportunity for soldiers to give women "pleasure" and so persuade them into a marriage that would diffuse Burmese "blood" and diminish the minority population (Apple, 1998, p. 44).

Rape as a weapon of war is an important topic to discuss because of its illustration of the manner of fighting in the civil and ethnic wars that are most common today. Theoretically, the issue of rape in warfare illustrates that geopolitical agency is often very aware of the multitude of geopolitical structures and their interrelationships. By targeting relatively weak individuals, an army can disrupt communities and cultures. However, such belief in the strategy, and its chances of success, are made possible by existing patriarchal structures that view women in particular and subordinate roles.

offspring bleak. Fatouma's goals are clear: "One day I hope I will be married . . . I hope I find a husband who will love me and my daughter" (p. A2). The structures of religious and ethnic tradition and honor make the accomplishment of these goals problematic. In the *New York Times* article, another rape victim reports how her husband has abandoned her and her children, one a product of rape, with little hope of remarrying and, hence, facing a lifetime of economic hardship and social isolation.

The two examples in this section illustrate that the geopolitical experience is immediate and personal. Individual geopolitical agents experience hopes and barriers within family, household, and local structures. The "all-seeing" global visions of academics such as George Modelski are not only abstractions compared to real experiences, but also identify structures seemingly removed from the immediate horror of rape and social rejection, or the destruction of one's home by bulldozers. It is geopolitical structures that help us understand the reasons behind the situations facing 'Adnan and Foutama, as well as the potential change for the better, however remote. Conflict between the people of Darfur and the Sudanese government, and also between Israel and the Palestinians, is framed within a metageography invoking the importance of collective national and ethnic identities, a them-and-us mentality requiring territorialized notions of belonging.

What of Modelski's model? Can we relate the geopolitical structure of cycles of world leadership to the complexity of individual actors? In some cases, I think so. On Monday February 14, 2005, the *New York Times* ran a report in its New York/region section, stories that address the local interests and issues of the New York readership. The report was the first in a series entitled "Deployed" (Semple, 2005, p. A20) and related the experiences of New Yorkers serving in the US armed forces. The byline of the story is Forward Operating Base Speicher, Iraq—an interesting geography for a "local" story. The story tells of the 42nd Infantry Division of the New York National Guard (akin to "reservists" or Territorial Army in Great Britain) who had been deployed with other National Guard and regular soldiers to be responsible for the security of four Iraqi provinces. The report states that it is the first time a National Guard division has been fully deployed in combat since the Korean War, and it is also believed to be the first occasion in which a National Guard unit has command over regular units in a combat situation.

Deployment is not a matter of a division, it is the experiences of individuals—in the case of the National Guard, people who are extracted from civilian jobs in a "local" place and put into combat across the globe. As the *New York Times* points out:

> The soldiers here are the men and women who deliver your mail, cook your entrees, answer your customer service calls and patrol your streets. They are truck drivers, students, social workers, youth counselors, cosmetologists, doctors, mechanics, firefighters, general contractors, a pool repairman, a tea salesman, the manager of a coffee shop and a lawyer from Long Island who said he represented "slumlords."
>
> (Semple, 2005, p. A20)

The article contains the requisite quotes from those who see their service as a matter of pride and national service at a time of war. Major Sal Abano, a 41-year-old chief information officer for an insurance company, in civilian life, is certainly proud. He also makes a geographic connection; "I think it's good for New Jersey to have a presence here" (Semple, 2005, p. A20). Sentiments echoed by Specialist Dominick Schoonmaker, a 34-year-old mechanic, who relates his service directly to the terrorist attacks of 9/11 and says, "You can't sit on the sidelines anymore" (Semple, 2005, p. A20).

Others in the division reflect "an undercurrent of regret and dread evident in many National Guard troops who acknowledge that they found themselves in a situation they neither imagined nor wished for." The brave confession of 43-year-old Staff Sergeant David C. DeMaio is: "When I got activated I cried like a baby." Given their recent and central roles as civilians in their communities, the deployment of the 42nd connects places in New York and New Jersey to places in Iraq. Sergeant Major James D. Rodgers is, in civilian life a letter carrier (postman) and says that his customers have kept in contact with him through e-mail and regular post: "They're very interested in how I'm doing out here," he says (Semple, 2005, p. A20).

In Chapter 2, we related the 2003 invasion of Iraq and the subsequent military occupation to Modelski's model of world leadership and, specifically, the need to place combat troops across the globe in the face of challenge to the leader's authority. Though one uses an abstract model to identify long-term trends and structures, the global politics of world leadership is enacted by individuals. These individuals have multiple goals and identities: simultaneously proud defenders of their country and scared family members hoping to return to the tranquility of civilian life. The pride in being a combat soldier stems from the role of the US as world leader and its representation as the global protector of "freedoms" and human rights. This representation is given greater resonance to individuals through narratives of national history.

The final quote from the article is the most significant though. The extended use of the National Guard has made a strong connection between communities in the United States and the combat zone, facilitated by e-mail and websites. The impact on communities and families as men and women fight abroad is made clear, and able to be communicated from combat zone to home front more quickly than in the past. Structures of family, community, military, nationalism, and global politics are tightly connected (Figure 8.2). In such a situation, the geopolitical representation of the war is a vital battlefield—if Sergeant Major Rodgers is able to report back to his community on what the combat experience is like, the need for his sacrifice to be seen as vital and moral is enhanced.

Modelski's model is an abstract simplification. The story of the National Guard places "people" within that model. By adding real and particular individuals into the picture, the descriptive nature of Modelski's model becomes less dominant, and instead we are forced to ask questions as to how and why actors operate within the structure. Noteworthy is the readiness by which the soldiers in this newspaper story say things that reflect the world leadership role of the United States. In other words, the language used by the US government in its world leadership role is echoed by individuals. Despite the pull of other identities (family, home, career, community, for example), identity and loyalty toward the United States nation-state *in its global role as world leader* is readily and unquestionably offered. However, the actions of the National Guard members are no means pre-determined. Their loyalties to family, for example, in addition to the way

Figure 8.2 Returning from war.

their actions are interpreted in their home communities (especially if opinion toward their military mission becomes negative) suggest that the level of their commitment may change. In other words, they are not automatons within a geopolitical structure, but complex agents who make decisions. It is well within the realms of possibility that people choose self and family over military service, for example. Indeed, the purpose of military recruiting campaigns is to ensure that people choose national service, while the modern military is very aware that long overseas deployment puts stress on the family and decreases retention.

Let us compare the geographic connectivity of the National Guard in a contemporary conflict with the experience of citizens drafted to serve in World War II. Fussell (1989, p. 288) notes the way war was reported and the self-censorship involved while writing letters home meant "a slice of actuality was off limits." In a statement that could be applied beyond the single case of World War II, he goes on to argue that war's "full dimensions are inaccessible to the ideological frameworks that we have inherited from the liberal era" (Fussell, 1989, p. 290). For Fussell the meaning and experience of war could never be comprehended except by those who witnessed combat.

The focus on the "messiness" of geopolitics shows that the reality and experience of conflict is very different from the simplicity and singular explanations provided by, say, Modelski's model or definitions of nationalism, etc. However, that does not mean that models and theoretical concepts are unimportant. The role of models and concepts is to simplify in order to understand key structures and processes. The trick is to realize that any given situation is the coming together of a variety of geopolitical agents and

Activity

Does the contemporary participation of citizen soldiers in relatively easy contact with the "folks" back home change the WWII separation of combatants and non-combatants and break down the barriers between battle experience and home?

Consider the ways contemporary welfare is represented. It has been argued that the digitization of war, including the computerized scenarios played out in the media and the US army's own website, produce a virtual war experience that is intense but quite false (Der Derian, 2001).

To consider the implications of the differences between now and World War II think back to the discussion of Orientalism in Chapter 4 and the way that the casualties suffered by the Iraqis have been removed from our understanding of the war in Iraq.

- In what sense does the way war is experienced "back home" depend upon images and understandings of "our soldiers" compared to the casualties of those they are fighting and the civilian population?

- Do we think differently of soldiers from our country if they are in intense gunfights or bombing people from a distance with high-tech weaponry?

- Do our opinions of war change if we see media coverage of our troops in action?

- Would public opinion regarding the war in Iraq, for example, change if we heard daily reports of enemy casualties?

---

structures operating at different scales. In other words, we can try and make sense of the "messiness" by first identifying the multiple structures and agents at work and, second, seeing how they interact in complementary and competing ways. The following case study is of the nationalist conflict in the disputed region of Jammu and Kashmir (Figure 8.3). As well as providing background to this particular conflict, the case study will attempt to show the interaction between geopolitical agents and structures, namely political parties, state officials, nationalist groups, and individuals.

Case study 8.1: persistent conflict in Jammu and Kashmir

Timeline

1752 Afghan warlord Ahmed Shah Durrani conquered Kashmir after the collapse of Mughal power.

1819 Due to the bloody nature of the Afghan rule, the Muslim majority surrendered all of their land to Punjabi Sikh King Ranjit Singh so that he could take over, beginning 27 years of Sikh rule.

1820 Ranjit Singh makes Gulab Singh Raja of the state of Jammu (starting here and continuing until the British takeover Gulab Singh begins to build a small empire, expanding his rule into the northwest regions now known as the Northern Territories). At the same time, Ranjit Singh gives the province of Poonch to Gulab Singh's brother, Dhyan Singh.

1846   Battle of Sabraon, British capture Lahore by defeating the successor of Ranjit Singh. In the resulting Treaty of Amritsar, the British award Gulab Singh with Kashmir (he did not send troops to help resist the British) and the title of Maharaja and the British retain "supreme control of the Valley" (Malik, 2002, p. 19).

1847   Gulab Singh dies and is succeeded by his son, Ranbir Singh.

1858   Beginning of the British Raj (occupation).

1885   Ranbir Singh dies and is succeeded by his son, Pratap Singh.

1906   All-India Muslim League founded to protect the rights of Muslims in British India.

1925   Pratap Singh dies with no heir. Hari Singh becomes Maharaja of Kashmir and Jammu. He is appointed by the British.

1931   April 29—Khutba (an important Islamic sermon) banned at a mosque in Jammu.

June 25—Pathan cook named Adbul Qadir makes "impromptu, highly inflammatory speech condemning Hindus in general and Hari Singh's rule in particular" (Malik, 2002, p. 34). He is immediately arrested.

July 4—Hindu police official, during an incident with a Muslim police constable, tore up Koran.

July 13—Some 7,000 gather for trial of Qadir. Police open fire, killing 21. Anti-Hindu riots occur all over Srinagar. This day is understood to be a turning point in relations between the Maharaja and his people.

Culmination of Kashmiri Muslim grievances.

1932   Formation of the All-India Jammu and Kashmir Muslim Conference Party, with Sheik Mohammad Abdullah as president of the new party. Initially this group united all of the Muslims, but divisions started occurring soon afterward.

1935–6   Poonch is integrated as part of Jammu and Kashmir (result of a lawsuit in the British Indian courts).

1939   Muslim Conference Party, headed by Sheik Abdullah, recognizes the need to secularize; change from Hindu vs. Muslim mindset to lower class vs. elite: name changed to the National Conference Party.

1941   Official reemergence of the Muslim Conference Party. The most notable figure was Yusuf Shah who had long-term ideological differences with Sheik Abdullah. Generally speaking, the National Conference was supported in Kashmir and the Muslim Conference in Jammu.

1942   Congress's "Quit India" campaign begins, goal of ending British rule of India.

1943–4 Numerous attempts to unite the Muslim Conference Party and the National Conference Party fail.

1944   The Muslim Conference Party's "New Kashmir" campaign emphasizes the desire to achieve rights for all, especially women.

1946   Yet another failed attempt to unite the two parties, motivated by the National Conference due to its declining popularity.

        May—"Quit Kashmir" campaign against Dogra rule, specifically Hari Singh, called the Treaty of Amritsar illegitimate.

1947   June 3—British announce plan to partition India.

        August 14–15—Ending their rule, British create the two separate independent states: Islamic Republic of Pakistan and India (the Radcliffe Boundary Commission is in charge of setting boundaries).

        ■ At the time of partition, the views of the people concerning Jammu and Kashmir fell into three general categories: Hindus (geographically concentrated in Jammu) wished continued rule of the Maharaja. The Muslim Conference members wished to be a part of an Islamic state (either Pakistan or independent) and the National Conference wished to join the secular Indian state (Muslims were the majority group in the Kashmir Valley and a large amount of Muslims were also in Jammu) (Malik, 2002, p. 64).

        ■ Mass killings (based on ethnicity and religion) and displacements were occurring at this time. Hindus and Sikhs moved eastward and Muslims were migrating westward.

        ■ During the British occupation there were areas that were formerly controlled by the British and areas, such as Jammu and Kashmir, where power was given, by the British, to another leader. At the time of partition, it was assumed that the latter would join either India or Pakistan, based on both geographic location and characteristics of the population. For most of the provinces this decision was clear for either one or both of the reasons listed above, whereas Kashmir and Jammu lay in between the two states, had a majority Muslim population, and were being ruled by a Hindu (Malik, 2002, p. 63).

        October 12—Statement issued by spokesman for Hari Singh stated the wish to remain independent and neutral—"the Switzerland of the east" (Malik, 2002, p. 64).

        ■ An uprising, beginning in Poonch, leads to the declaration of an independent Azad Kashmir by its Muslim majority.

        October 26—Faced with incoming Pakistani tribal troops, Hari Singh is forced to sign the Instrument of Accession to India. India sends troops to secure the area.

■ The accession is set to include Jammu and Kashmir as a formal part of India upon ratification.

1948   March 5—Interim government formed with Sheik Abdullah as Prime Minister (at this point, Hari Singh still holds title of Maharaja but with little to no power).

May—Indo-Pakistan war begins, when Pakistan sends its official troops into Kashmir.

1949   January 1—With UN intervention (UNCIP = UN Commission on India and Pakistan) a ceasefire stops the war from spreading into the rest of India and Pakistan.

■ There were an approximately even number of casualties on both sides (1,500 for India and 1,500 for Pakistan).

January 27—The official ceasefire line is declared and remains until 1965.

■ The region was then separated into three different administrative parts: the Northern Areas (controlled by Pakistan), Azad Kashmir (independent in theory), and the rest which was controlled by Indian troops.

1952   Dogras' hereditary position is officially abolished. Relations with India through this entire period are ambiguous because accession was still not ratified.

June 24—Delhi Agreement—Jammu and Kashmir are part of India, but with a higher level of autonomy (but by the following year, Sheik Abdullah was involved in conversations about sovereignty for Jammu and Kashmir with US and UN officials). Hindus opposed the Delhi Agreement because it meant that they would not be protected by New Delhi from Kashmiri Muslim rule.

1953   August 9—New Delhi arrests and replaces Abdullah. Bakshi Ghulam Mohammed is sworn in as new prime minister. Protests occur during the next few weeks. Bakshi has little support from the people, needs New Delhi to keep him in power, but Kashmir experiences a period of stability nonetheless.

October 5—Legal framework is laid for formal accession to India and increased power of New Delhi in Kashmiri affairs.

1954   Pakistan signs a military aid agreement with the United States.

1955   Soviet leaders visit India, making a trip to Srinagar.

Plebiscite Front (aka Action Committee) emerges, an opposition group formed by Mirza Afgal Beg and supported by Sheik Abdullah (anti-Bakshi and New Dehli). This group's goals were less centered around autonomy and more geared toward proving the Instrument of Accession invalid due to the fact that Jammu and Kashmir were only to become part of India after a popular referendum that never occurred.

1956–7   Following years of friendly Indo-China relations, China begins to build a military highway in disputed territory, Aksai China.

1957   January 26—New constitution of Jammu and Kashmir takes effect reaffirming the accession to India. USSR exercised its UN Security Council veto for the first of what would come many times during a discussion of Kashmir initiated by Pakistan.

1959   India sends border patrols into the area under dispute with China.

1962   Small fights break out between India and China beginning the border war (47,000 square miles of disputed land).

October 4—Bakshi resigns and Revenue Minister Khwaja Shamsuddin is sworn in.

October 10–November 20—Significant fighting between Indian and Chinese forces. The Chinese People's Liberation Army was well prepared for the fighting in the Himalayas—they were in warm, padded uniforms and had previously fought Tibet in the same climate (14,000–16,000 feet altitude). The Indian army had a small budget and ill-prepared troops.

November 21—China, after accomplishing all goals of land attainment, declares a ceasefire. "Following the ceasefire, China kept most of her claim in Aksai China but gave India virtually all of India's claim in the North East Frontier Agency—about 70% of the disputed land!" (Calvin, 1984). According to official Indian reports the number of Indian casualties was 1,383 troops killed, 3,968 captured, and 1,696 missing. The data were never released for Chinese casualties.

1964   The Action Committee splits and the Awami Action Committee forms.

May 23—Abdullah travels to Pakistan for negotiations (after being jailed by the Indian government numerous times he became a hero in Pakistan).

May 27—Prime Minister Nehru dies—relations between New Delhi and Abdullah decline rapidly—negotiations with Pakistan fail.

1965   Indo-Pakistan war.

July—Pakistan begins sending troops, anticipating support from Kashmiris.

September 6—India's counter-attack crosses the border into Pakistani Punjab.

September 23—UN-mediated ceasefire. By the time of the ceasefire, Pakistan suffered approximately 3,800 casualties while India suffered approximately 3,000 (Indian Express Group, 2001). National Conference Party changed name to the Pradesh Congress Party (extension of New Delhi government).

1966   January 10—Tashkent Declaration, result of Russian-mediated peace talks. India and Pakistan move back to pre-war borders, repatriate POWs, and re-establish diplomatic relations.

1971   Indo-Pakistan war. Originally beginning with a civil war in East Pakistan, which becomes Bangladesh by the end of the war; seen as the liberation of Bangladesh from the Indian perspective. In Kashmir, The Plebiscite Front is banned. Abdullah is externed from the state.

July 2—The Simla Pact is signed by both sides. They agree to respect the line of control until further resolutions are made.

1975   Accord between Abdullah and Indira Gandhi, prime minister of India. India sees it as firming the union. Abdullah sees it as protecting Kashmir's special status. He returns to power.

1981   Farooq Abdullah, the sheik's son, takes over office.

1982   Sheik Abdullah dies.

1984   Farooq is dismissed in a "drawing room dismissal" engineered by Indira Gandhi. Protest ensues. Farooq is replaced by G.M. Shah, who is an unpopular ruler.

1986   Rajiv Gandhi's, India's new prime minister, government reinstates Farooq as chief minister—less popular now in Kashmir because of his collaboration with India.

1987   Insurgency in Kashmir gains momentum from this time on. Farooq blames unemployment, especially of the educated, with about 40–50,000 unemployed graduates. Others point to rigged election forcing a resort to armed struggle. India responds by intensifying its security actions in the region.

1990   Central rule imposed on Jammu and Kashmir (with dissolving of the Jammu and Kashmir State Assembly. Between 1989 and 1990, 140,000 Hindus leave the Kashmir valley—they go to refugee camps in Jammu.

1992   Operation Tiger (followed by Operation Shiva) carried out by Indian security forces. These security operations have led to allegations of widespread killings and other atrocities.

1998   Both India and Pakistan begin nuclear testing.

1999   Indian and Pakistani militaries clash in Kargil.

2001   December—The Prevention of Terrorism Bill (POTB) is passed: "a repressive piece of legislation that could be used to justify considerable human rights abuses by the government of India, especially in Kashmir, where India is fighting a counterinsurgency war" (Podur, 2002).

Accurate, reliable information concerning the amount of casualties since the beginning of armed conflict in Kashmir is impossible to obtain. Official handouts give the following information from 1990–9: 9,123 members of armed opposition groups; 6,673 victims of armed opposition groups; 2,477 civilians at the hands of Indian security forces and 1,593 security personnel have been killed. However, the Institute of Kashmir Studies, a research center, has estimated the number of 40,000–50,000 deaths since 1989/90 (all information taken from the 1999 report from Amnesty International, "India," 2001, pp. 8–9). Since 2001, tensions between India and Pakistan have waxed

Figure 8.3 Historical roots of conflict in Kashmir.

and waned. In early 2005, there were signs of cross-boundary cooperation that may be interpreted as peaceful overtures, and an earthquake in the region in October 2005 resulted in promises of cross-border cooperation. However, the situation would change dramatically if terrorist attacks in India were resumed, as Indian politicians and the public are quick to claim Pakistani sponsorship of the Kashmiri militants.

Geopolitical agency in Jammu and Kashmir

The timeline emphasizes the actions of the Indian and Pakistani governments, and different national groups. If we explore the viewpoints of some of the major geopolitical agents, we will see not only the major points of contention, but also how different geopolitical structures combine to provide a context for agency. Indeed, the purpose of this case study is to emphasize how different geopolitical structures and agents interact. The goal is to show the complexity of geopolitical conflicts. The perspective from the Indian government has been consistent, identifying the violence in Kashmir as "an internal affair of India" (Srivastava, 2001, p. 95). At the beginning, an attempt was made to secure Kashmir as a part of the Indian state through Nehru's relationship with Sheik Abdullah and the National Conference in the hope of gaining support from all Kashmiri Muslims. In hindsight, we can see this was a major misjudgment of popular opinion. Consequently, India has had to use violence in its goal of maintaining Kashmir within the boundaries of the Indian state. In the language we introduced in Chapter 5, the rhetoric of the Indian government, has framed the conflict as the maintenance of the Indian state in the face of what they classify as insurgency.

However, it is wrong to see India's policy as singular or uncontested. Different Indian political parties have addressed Kashmir in their platforms. The following political positions reflect the stance of the parties in the 2004 Indian elections. The Bharatiya Janata Party (BJP) or Hindu Nationalist Party spoke of facilitating dialogue with Pakistan— the pressure on the party was to portray itself as being able to pursue a national agenda in a multicultural society. Especially, the party spoke of secularization in order that it could claim the ability to work with Muslims in Kashmir and across India in order to integrate Kashmir as a part of India (Upadhyay, 2000). The religious and ethnic identities that had provided the BJP with its electoral success had to be negotiated by the party, in an attempt to make peace across national and religious lines.

Another major Indian party, the Indian National Congress, emphasized security issues, especially the threat of terrorism, in its political campaign. The party's website juxtaposed the metageography of terrorist networks with a notion of a harmonious multinational state: "Indian National Congress will forcefully resist all attempts at using the issue of cross-border terrorism to polarise our society on religious and communal lines" (Sankalp, 2003). Defense of state boundaries was used by the party to make a claim for broad national support.

The conflict in Kashmir extended beyond the immediate region. The tension between Hindus and Muslims was focused upon geographic locations religiously significant to both groups. In 2002, 53 Hindus were killed in a terrorist attack on a train that was returning them from a religious voyage to Ayodhya. They had started to plan the erection of a Hindu temple at this site, which is of importance to both Muslims and Hindus (Lineback, 2002, p. 1). The bloodiest example of this type of conflict in India occurred in Mumbai (Bombay). Here riots broke out after Hindus destroyed a Muslim temple that they believed to be built upon the birthplace of their god, Rama. Eight hundred people died during these riots (Lineback, 2002, p. 1). Conflicts throughout India continue to occur. It is primarily a Hindu country but has the second highest number of Muslims (136 million, 14 percent of the population) after Indonesia (Lineback, 2002). The nature of these conflicts illustrates the reflexive consideration of religion; in other words, religious loyalties are entwined with nationalist struggles. Religious identity reinforces conflict over the structure of the nation-state, and simultaneously religious organizations and beliefs are reinforced within a context of nationalist struggle (Stump, 2005).

The official position of Pakistan takes a different approach to the conflict by trying to make a moral argument of national self-determination. On the official website of the Islamic Republic of Pakistan, amongst nine topics (including "Government," "Country Profile," "Economy") is the topic "Kashmir." In the "FAQs" portion on Kashmir, the Pakistani government states through the very first answer that Kashmir is different from other territorial disputes in that

> the territory involved is a whole country. . . . Here the matter is not one of placing a few hundred square miles on one side or the other of an international frontier and thus settling a boundary conflict. It is a matter of the disposition of a country through the same process by which the two contestants, the Indian Union and Pakistan themselves emerged as independent states—the process of establishing sovereignties on the basis of popular consent.
> (Islamic Republic of Pakistan, 2004)

Furthermore the website continues to explain that neither the Maharaja's signature on the Instrument of Accession nor the support of Sheik Abdullah and the National Conference legitimize the accession to India. This is based on the argument that these acts were not done with the consent of the Kashmiri people. The government of Pakistan's argument resorts to the ideology of nationalism, that the will of a nation demands a state.

The conflict is not just about one state versus another, or even a singular nationalist claim, though. Religious identity, ethnicity, age, and gender are all important structures that combine in different ways. People in the Muslim community have experienced severe treatment from the ever-present Indian security forces. "What unites disparate ideologies and programmes as well as ordinary people is a common enemy—the security forces" (Women's Initiative, 2002, p. 90). Due to mistreatment, a feeling of favoritism of the Indian government toward Hindus, and unfair elections, militant groups (known as *tanzeems*) have arisen.

"The main political division among Kashmir Muslims now is between those wishing to accede to Pakistan and those wanting an independent state" (Malik, 2002, p. 357). The similarity among all of the Muslims, however, is the desire to be free from India. *Tanzeems* are responsible for murders, rapes, and kidnappings of both Hindus and fellow Muslims. Because *tanzeems* are plentiful and uncoordinated, rivalries result that spur violence between groups. Fundamentalist groups also attack fellow Muslims that act in a way that violates their ideologies. For example, a teenager reports that his father was murdered because he consumed alcohol: "The Hizbul Mujahidden had warned him about drinking but when he didn't care they killed him" (Chhachhi, 2002, p. 201).

In 1987, following elections that were thought to be fixed (given the very poor performance of the Muslim United Front in the elections), the youth of Kashmir began to protest and many were arrested. Disaffected youth, who believed they were perse-cuted for their religious beliefs and ethnicity, were sought by the Inter-Services Intelligence (ISI—the Pakistani Intelligence Service) who promised "arms and training to these 'boys' to launch armed struggles against India" (Santhanam et al., 2003). These recruits came mostly from the Islamic Students League (ISL). In the mid-1990s, due to the enormous amount of casualties within the *tanzeems*, there was a "drying up of young Kashmiri recruits. . . . School dropouts and rowdy elements began to dominate. . . . Rape became common in the Valley while innocent civilians were murdered on the suspicion that they were 'informers'" (Santhanam *et al.*, 2003, p. 28). According to a four-member all-woman team who set out to assess the situation in 1994:

> Many people reported the recruitment of thousands of Kashmiri youth from poor families. . . . Someone remarked, "The sons of the rich in India and Pakistan go to America to study, for better opportunities. Our boys go out to learn how to use the gun. The power brokers are not interested in stopping the war, their children are not being sacrificed."
>
> (Women's Initiative, 2002, pp. 89–90)

Lack of opportunity for young men, oppression by Indian security forces, a willing sponsor, and nationalist and religious ideology combined to fuel the *tanzeems*.

The *tanzeem* itself could become the most meaningful geopolitical structure, promoting disputes and violence between groups despite claims of a common cause. Brief description of four *tanzeems* shows the mixture of shared and divergent goals. *Hizbul Mujahideen* (HUM) emerged as an important *tanzeem*, headquartered in Srinagar. It is sponsored by the Pakistan government, ISI, and Jamaat-e-Islami, a political party in Pakistan. The objectives of the HUM are to succeed from India via armed combat and to merge with Pakistan. In 1990, Jamaat-e-Islami and the ISI took control of HUM. Now HUM is considered the militant wing of *J & K Jamaat-e-Islami* (JKJEI). JKJEI is closely linked with the *Jamaat-e-Islami* (JEI) *tanzeem* in Pakistan. This group was more politically oriented and won seats in the 1987 State Assembly elections (Santhanam *et al.*, 2003, p. 154). This *tanzeem* obviously shares the same objectives as the HUM, but seeks different means. The Jammu and Kashmir Liberation Front (JKLF) differs from HUM and JKJEI by its goal of an independent, united Jammu and Kashmir (including Pakistani occupied Kashmir and the Northern Territories). This group formed in 1988 (when the ISI was easily recruiting angered students and creating many new *tanzeems*) and has its headquarters in Srinagar. Finally, the Jammu and Kashmir People's Conference (JKPC) is less radical in nature with the objective of greater autonomy for the state of Jammu and Kashmir under the Indian Constitution. Two points should be taken from the diversity of the *tanzeems*. First, the protagonists in a conflict are rarely unified, and so it is wrong to view a particular cause or issue as singular. Second, the variety of geopolitical structures produced different goals and identities that were mutually reinforcing.

The creation of ethnic difference is also evident in this dispute. A conflict over the location of an international boundary fermented a conflict in which group identity became significant. It is estimated that about 400,000 Kashmiri Pandits (a sect of Hindus with ancestral ties to Kashmir) were forced from there homes between 1989 and 1991. The fears of a Pandit doctor facing a crowd of Hindus outside her house illustrates how cultural conflict was created over time: "Many of the young men in the crowd were boys I had delivered at the hospital! And here they were now shouting for my blood" (Raina, 2002, p. 179). The status of Pandits has changed too as they have been forced to become refugees:

> While the "refugees" were earlier welcomed and given assistance, local people have now begun blaming them for being the cause of all problems, ranging from typical urban infrastructure shortages of water and transport, to unemployment, . . . increased state violence, militant attacks, sexual harassment, etc.
>
> (Dewan, 2002, p. 154)

Prior to the recent violence, Pandits and Muslims lived side by side without any problems. One women recalls that "before the Kashmir issue [her] friends from that region were just Kashmiris; they were not seen as Muslims or Pandits" (Dewan, 2002, p. 149). Now the situation is quite different and Pandits' wishes for the fate of Kashmir differ greatly from those of the Kashmiri Muslims. "They want their own exclusive 'state' within the Valley—Panun Kashmir. This would be a region or state within India, autonomous both from central government and Kashmiri Muslim control" (Malik, 2002, p. 358). In other words, as conflict creates group identity and ethnic violence the desire for a state for one's own group is seen to be imperative, and the geopolitical structure of a world of nation-states is reinforced.

To end our discussion of this conflict, a consideration of gender illuminates overarching or dominant geopolitical structures, as well as the cracks in their foundations. For the most part, the suffering endured by Muslim women on an individual level in the conflict is practically identical to the situation facing Pandit women who are normally seen to be on the other "side." The common threat of rape (see Box 8.1) illustrates how the structure of patriarchy transcends nationalist and religious conflicts. The perspective of women is also able to stress comprehension of shared values and seek compromise and fusion over conflict and hierarchy. The sentiment of most women is for peace based upon shared experiences. As one Pandit woman said:

> It was after years that we had all gotten together at a marriage—all of us women—Pandit, Muslim, Sardarnis. It was almost like the old days . . . We laughed and danced late into the night. Then, as we prepared to go to sleep, I heard some of the Muslim women whispering among themselves in the next room: "It's been such a lovely evening. It is true, isn't it, that a garden is only a garden of any worth when there are many kinds of flowers gracing it."
>
> (Chhachhi, 2002, p. 207)

However, not all women are united by feminist beliefs that negate the geopolitics of nationalism. A minority of women in the region see their primary role to be within nationalist movements. For example, *Khawateen Markaaz*, originally an organization that carried out social work for Muslims in Kashmir, joined the *Azaadi* Movement in 1990. This group wishes for an independent Kashmir and believes:

> Kashmir is occupied by both India and Pakistan. We are Kashmiri women. We are committed to independent Kashmir. We respect all religions. We are not fundamentalists. People of all religions will live side by side. Kashmiri pandits should come back here, this is their motherland.
>
> (Women's Initiative, 2002, p. 86)

On the other hand, *Dukhtaran-e-Milat*, begun in 1980, wishes that Kashmir become part of Pakistan. The movement uses the terms Hindustan and *jihad* and its leader says that "if the men make a pact with Hindustan, women of *Dukhtaran-e-Milat* will pick up the gun even against our own men if need be" (Women's Initiative, 2002, p. 87). Clearly, the imperatives of nationalism are more important than submitting to traditional gender roles here. Ironically, the motivation is far from progressive though, as men are to be challenged only if they are seen to be nationalist appeasers.

Many women, the majority who are not themselves a part of militant activity, accept the supporting role to men in their lives who join *tanzeems*. In other words, structures of patriarchy implicate women in the conflict through their subordinate relationship to husbands. Two women from Bandipora express their acceptance of family members taking part in militant activity: "I knew my husband was a militant. I knew that some day he would be killed. I grieve, but I do not complain" (Chhachhi, 2002, p. 194). Regarding her son, another woman said: "The child of a freedom fighter will be a freedom fighter" (Chhachhi, 2002, p. 194).

The case study of Kashmir has emphasized the diversity of geopolitical agents within a particular conflict and the intersection of a number of geopolitical structures. In our framework we note that geopolitical agents have opportunities and constraints set by geopolitical structures. If agents and structures are both multiple, then the choices made by geopolitical agents and the identities that come to the fore will be complex or "messy." However, the "messiness" is a function of the many geopolitical structures that interact to form particular geopolitical contexts. Moreover, geopolitical agents have, well, agency, or the ability to make choices—such as the Kashmiri women who either rejected the language of nationalism or adopted it. By emphasizing "messiness" in this final chapter two things should stand out. First, no geopolitical conflict is simple—there are divisions within the antagonists, or many different struggles (gender, race, religion, etc.) are in play within what is often reported as a "one-issue" situation. Second, you can understand the complexity by identifying the different structures that are operating, and noting the way they intersect. As a result, an attempt can be made to identify and understand the options (or lack of them) available to the different agents.

Messiness and structure

The "messiness" of geopolitics is a term aimed at emphasizing the intersection of gender, generation, ethnicity, and religion with structures that are more commonly associated with geopolitics, namely state-versus-state competition and national identity. However, geographies of the global projection of power, inter-state conflict, and nationalism are not to be ignored. At the time of writing, three stories in one edition of the *New York Times* (February 18, 2005) stressed these types of conflict. First, the front page of the "Business Today" section spoke of the increased demand for oil from India and China as their economies grow. The story reports that China's oil imports grew by 31 percent between 1994 and 2004, and India's by 8 percent. Traditional suppliers to the United States, such as Saudi Arabia, recognized the diversity of customers and have responded

Activity

Find a news magazine such as the *Economist, Atlantic Monthly*, the *New Yorker, Time*, or one of the color supplements of the Sunday newspapers. These magazines usually carry longer stories on current conflicts than the daily newspapers and include interviews with the participants and victims.

Explore an article of your choice and use the interviews and descriptions of the participant's circumstances to identify different structures and how they interact.

Do the interviews show divisions within particular groups or agents, such as political parties, ethnic groups, etc.? In other words, does the article exemplify how geopolitical agents are not singular?

If you are in a class or other group setting, you could do this project with someone else and explore the same conflict using different media sources. This will not only help you in identifying more structures and types of agency, but you may also consider how different media outlets emphasize different structures and types of agency over others. For example, were political parties and state ministries or departments emphasized in one source while protest groups, women's groups, and other social groups emphasized in another?

by increasing their output. However, India and China are also aware that they are latecomers to the world oil markets and have responded by tapping new areas of supply in countries that are seen as geopolitically isolated, namely Sudan and Myanmar/Burma. Referring back to the beginning of Chapter 4, it was Lenin's belief that competition for resources was the source of the twentieth century's global wars. It is important to recognize the geopolitics of the competition for the earth's resources (especially oil and water), and the conflicts between states that can result (Box 8.2).

Within our framework of the geopolitics of world leadership the second story is clearly connected to the first. Interestingly, the *New York Times* made no such connection. The story, on page A6, reports growing awareness by the US military of the expansion of the Chinese navy and its ability to project power across the globe. Defense Secretary Donald H. Rumsfeld is reported as saying, "They're [the Chinese] increasingly moving their navy further distances from their shores in various types of exercises and activities. And that's a reality." In the report the increased strength of China's navy is linked to the potential conflict over Taiwan, but the increased geographic scope of China's navy may also be linked to its global pursuit of oil supplies. Furthermore, the report notes US anger at the EU's plan to sell weapon's technology to China. Referring back to Modelski's model, the actions of the EU may be interpreted as delegitimation, while the build up of China's navy and the potential for fighting in the Taiwan Strait and beyond points to the possibility of subsequent phases in the cycle: deconcentration and global war.

The third story refers to developments in post-war Iraq, and especially the political horse-trading that occurred after the elections of January 2005. The strong showing of the Iraqi Kurds in the election, second behind a Shi'ite alliance put them in a very strong position, as discussions were underway to put together a ruling coalition. The Kurds

Box 8.2 Geopolitics of nature

Classic geopolitics was concerned with nature in a particular manner; access and control to resources. Resources remain a central calculation in contemporary geopolitics. Indeed, many recent publications are eager to point out the role of oil in the current deployment of US troops in the Middle East and Central Asia. Another way in which nature enters contemporary geopolitics is the definition "failed states" given to countries that do not have stable governmental institutions to facilitate the extraction of resources. These two approaches to nature illustrate the feminist and Marxist critiques of the geopolitical vision: geopolitics is about gaining access, control and domination through the division of territory.

In addition to questions over the control of natural resources, the awareness of the fragility of the global ecosystem has also provoked transnational ecological geopolitics. On the one hand, social movements such as the anti-globalization movement discussed in Box 7.1 and NGOs such as Greenpeace are geopolitical agents with their own agenda that is not limited to national interest. On the other hand, negotiations between states over global warming, for example, reflect state calculations about economic pros and cons, but within a context of growing awareness of a common global problem.

The tension between interstate competition for natural resources and the geopolitical agency of social movements illustrates the way in which very different geopolitical agents combine to set agendas and make change.

For further reading about the geopolitics of resource control see Klare (2004). For a selection of readings about the geopolitics of environmentalism see the collection of essays on environmental geopolitics in *The Geopolitics Reader* (Ó Tuathail *et al.*, 1998).

demands centered upon increased autonomy for the Kurdish area of northern Iraq. However, the demands read as a manifesto for an independent Kurdistan and included: ownership over the regions oil reserves and control of the revenues; authority to maintain a 100,000 strong armed militia; ministries that would parallel those in Baghdad; and authority over fiscal policy. The dominant geopolitics in this case is that of nationalism, the pressure for an independent Kurdistan. Also, conflict over the structures of the nation and state in this case are part of the broader struggle of the United States to achieve its goals in the Arab world. In what may also be interpreted as evidence of delegitimation, as defined by Modelski's model, the election results as a whole, the hostility of segments of the Iraqi population, and the lack of acquiescence toward US goals by many others, suggest that both the coercive and integrative power of the world leader are being challenged. In October 2005, by way of a referendum, the people of Iraq approved a new constitution. However, the negotiations prior to the vote between Iraqi parties over the form of the constitution show the potential for future conflict.

The Kurds wanted a federalist constitution giving them, and the Shia Muslims in the south of the country, greater autonomy, but this was being resisted by the minority Sunni Muslims in the central region. The Sunnis saw federalism as a probable step toward the disintegration of Iraq, and their consequent exclusion from oil revenues. Iraq's oil is located in the north and south of the country. The agency of the Kurds in their attempt to create a Kurdistan focuses on one geopolitical structure, the state, in a way that is both facilitated by, and has an impact upon, the geopolitical structure of world leadership.

By emphasizing how, on the one hand, Chinese agency is based upon a desire for economic growth and growing participation in world trade on the one hand, and increasing military strength on the other the complexity, or multiple motivations, of geopolitical agency is evident. The agency of the Kurds illustrates that identity is mobilized within a suite or hierarchy of structures, and the intended implications toward one set of structures may have unintended consequences on other, non-targeted, structures. In other words, messiness and structures go hand in hand.

Conclusion and prologue

A book such as this has no definitive conclusion. The book's task is to let the reader initiate inquiry into geopolitics and not to provide things that are "known." The case studies are included to provide background to what have proved to be persistent conflicts that could intensify and expand. Knowledge of these actual conflicts is necessary to understand contemporary geopolitics in two senses: the basic "what is happening/where is Chechnya?" sense and as a way to exemplify the manner in which geopolitical structures and agents interact. In the first sense, the case studies provide a stepping-stone toward a knowledge that will steadily expand as you continue to explore and engage current affairs. In the second sense, the case studies are my attempt to talk you through some actual conflicts with reference to the framework of structures and agents— they are an exercise that I hope will facilitate your ability to analyze future geopolitical situations.

If I have one goal with this book it is to make you informed and active participants in geopolitics. In the most everyday sense, I hope that working through this book allows you to critique what you see and hear in the media. When an "expert" is put in front of the cameras or framed on the opinion pages do not be in awe of them, but use the perspective and knowledge you have gained form this book to question their assumptions, the way they approach the conflict and so limit the questions that are asked, wonder what someone from another national, gender, class, racial, or religious political perspective would say instead. To do this, the first thing is to tease out *all* the geopolitical structures and agents that are involved in the conflict and, hence, be aware of what the expert is *not* discussing. The next step is to construct a fuller picture than the expert will deliver by integrating the role of the excluded agents and structures.

My other intention for writing this book is to act as a guide to participating in geopolitics, but I am aware that this is a pretentious claim, so please let me qualify the statement. I hope that one of the lessons from this book is clear: we are all geopoliticians, we participate on a daily basis. We recreate our own national and state structures by simple acts of reading a "national" newspaper that is organized to talk about "them" in the international section as opposed to "us" in the politics, sports, and weather sections (Billig, 1995). We carry around images of other countries and conflicts that are based upon popular representations of geopolitics, which in turn influence our approval of or opposition to foreign policy. Being aware of the structures of global interstate interaction, and nationalism, may, at the very least, allow for more reflection when one is asked to act in the name of the "common sense" that such structures inspire—a common sense that feminists will be eager to point out revolves around hierarchy, difference, and violent competition. What are the structures and notions of "normal" behavior underlying *Dulce et Decorum Est pro Patria Mori*?

For many, participation in geopolitics is much more than the passive reconstruction of structures that are remote and somewhat intangible. Career paths may well lead to direct involvement: teaching in the United States I am responsible for the education of many young adults who have already begun serving in the armed forces or wish to pursue careers in intelligence agencies or as part of the Department of Homeland Security. For many of the current generation of university students, political awareness was initiated on September 11, 2001. Their sense of geopolitics is very much molded by the language of the War on Terrorism.

Participation in geopolitics is also a matter of questioning and challenging the "common sense" assumptions generated by the geopolitical structures in general (difference, conflict, etc.) as well as by the representations and actions of key geopolitical agents, the US and British governments for example. Protest, dissent, questioning, are also evident among the students I teach—disaffection with both the persistent structures as well as specific government actions are also common viewpoints that produce their own actions.

I am in no position to be judgmental about the geopolitical actions that others take. The message that I want to end with is that agency is constrained and enabled by structures. You have choices within structures—knowing the structure makes for a more informed strategy—whether that is within a family, neighborhood, business, social movement, or state. The same awareness may be applied when interpreting current events. The decisions made by the governments of Iran and North Korea, for example, may be portrayed as irrational and unnecessarily aggressive by Western governments. But it is your task to see them as agency within structural settings of global, regional, and inter-state politics. As Robert McNamara advised in the excellent documentary *The Fog of War*, Lesson Number One is to empathize with your enemy. Knowing the structural context of other geopolitical agents is a means to knowing their fears, concerns, and goals. Such knowledge of geopolitics is an avenue to empathy and understanding that will, I hope, be a pathway to a more peaceful world.

*

To finish, one may be more poetic in considering structures and agents:

> The world is big. Some people are unable to comprehend that simple fact. They
> want the world on their own terms, its peoples just like them and their friends,
> its places like the manicured little patch on which they live. But this is a foolish
> and blind wish. Diversity is not an abnormality but the very reality of our
> planet. The human world manifests the same reality and will not seek our per-
> mission to celebrate itself in the magnificence of its endless varieties. Civility
> is a sensible attribute in this kind of world we have; narrowness of heart and
> mind is not.
>
> (Chinua Achebe, Bates College commencement
> address, May 27, 1996)

Further reading

The readings listed at the end of Chapter 1 as more detailed and sophisticated investigations of geopolitics should be reviewed. They will provide different interpretations and topical concentrations that will be accessible after reading this book.

Allen, B. (1996) *Rape Warfare: The Hidden Genocide in Bosnia-Herzegovina and Croatia*, Minneapolis, MN: University of Minnesota Press.

A discussion of the role and meaning of rape in warfare, with a detailed case study.

Bose, S. (2003) *Kashmir: Roots of Conflict, Paths to Peace*, Cambridge, MA: Harvard University Press.

Provides an understanding of a long running conflict that has broader regional implications.

Stump, R.W. (2005) "Religion and the Geographies of War" in Flint, C. (ed.) *The Geography of War and Peace*, Oxford: Oxford University Press, pp. 144–73.

Provides a framework for identifying and interpreting the role of religion in conflict.

REFERENCES

Abraham, D. (1986) *The Collapse of the Weimar Republic*, second edition, New York: Holmes and Meier.

Agamben, G. (1995) *Homo Sacer: Sovereign Power and Bare Life*, translated by D. Heller-Roaxen, Stanford, CA: Stanford University Press.

Agnew, J. (1987) *Place and Politics*, London: Allen & Unwin.

Agnew, J. (2002) *Making Political Geography*, London: Arnold.

Ahmad, E. (2000) *Confronting Empire*, Cambridge: South End Press.

Alexander, Y. and Swetnam, M.S. (2001) *Usama bin Laden's al Qaida: Profile of a Terrorist Network*, Ardsley, NY: Transnational Publishers.

Allen, B. (1996) *Rape Warfare: The Hidden Genocide in Bosnia-Herzegovina and Croatia*, Minneapolis, MN: University of Minnesota Press.

American Forces Press Service webpage (July 11, 2004) "About Operation to Blue to Green," www.usmilitary.about.com/od/armyjoin/a/bluetogreen_p.htm. Accessed November 1, 2005.

Amnesty International (1999) "India: If They Are All Dead, Tell Us: 'Disappearances' in Jammu and Kashmir," March, 2 1999. www.web.amnesty.org/library/Index/ENGASA 200021999?open&of=ENG-IND. Accessed March 6, 2006.

Amnesty International (2004) "Israel and the Occupied Territories Under the Rubble: House Demolition and Destruction of Land and Property," May 18, 2004. www.web.amnesty. org/library/index/engmde/150332004. Accessed February 5, 2005.

Anderson, B. (1991) *Imagined Communities*, second edition, London: Verso.

Anderson, M. (1996) *Frontiers: Territory and State Formation in the Modern World*, Oxford: Polity Press.

Appadurai, A. (1991) "Global Ethnoscapes: Notes and Queries for a Transnational Anthropology" in Fox, R.G. (ed.) *Recapturing Anthropology*, Santa Fe, NM: School of American Research Press.

Apple, B. (1998) *School for Rape: The Burmese Military and Sexual Violence*, United States and Thailand: Earth Rights International.

Armstrong, K. (2000) *The Battle for God: A History of Fundamentalism*, New York: Ballantine Books.

Associated Press (2003) "Russian Soldiers' Mothers Fight for Young Draftees," January 14, 2003. www.hrvc.net/news/16b-1–2003.html. Accessed July 14, 2004.

Bacevich, A.J. (2005) *The New American Militarism: How Americans are Seduced by War*, Oxford and New York: Oxford University Press.

BBC (2004a) "Key findings on al-Qaeda," BBC News, June 16, 2004. www.news.bbc.co.uk/ 2/hi/americas/3813453.stm. Accessed August 25, 2004.

BBC (2004b) "Al-Qaeda's origins and links," BBC News, July 20, 2004. www.news.bbc. co.uk/go/pr/-/2/hi/middle_east/1670089.stm. Accessed August 20, 2004.

Beaverstock, J.V., Smith, R.G., and Taylor, P.J. (2000) "World-City Network: A New Meta-geography?," *Annals of the Association of American Geographers* 90, pp. 123–34.

Bernstein, D. and Kean, L. (1998) "Ethnic Cleansing: Rape as a Weapon of War in Burma," Burma Forum Los Angeles. www.burmaforumla.org/women/rape.htm. Accessed July 18, 2004.

Billig, M. (1995) *Banal Nationalism*, Thousand Oaks, CA: Sage Publications.

Boyd, A. (1998) *An Atlas of World Affairs*, tenth edition, London: Routledge.

Bradsher, K. (2005) "Two Big Appetites Take Seats at the Oil Table," *The New York Times*, February 18, 2005, p. C1.

Bregman, A. and El-Tahri, J. (2000) *Israel and the Arabs: An Eyewitness Account of War and Peace in the Middle East*, New York: TV Books.

Brogan, P. (1990) *The Fighting Never Stopped*, New York: Vintage Books.

B'TSELEM (2002) "Israel's Policy of House Demolition and Destruction of Agricultural Land in the Gaza Strip," www.btselem.org/English/Publications/Summaries/Policy_of_Destruction.asp. Accessed November, 15 2004.

Calvin, J.B. (1984) "The China India Border War," GlobalSecurity.org. Written: 1984. www.globalsecurity.org/military/library/report/1984/CJB.htm. Accessed June 1, 2004.

Chhachhi, S. (2002) "Finding Face: Images of Women from the Kashmir Valley" in Butalia, U. (ed.) *Speaking Peace: Women's Voices from Kashmir*, London and New York: Zed Books, pp. 189–225.

Chiozza, G. (2002) "Is There a Clash of Civilizations? Evidence from Patterns of International Conflict Involvement 1946–97," *Journal of Peace Research* 39, pp. 711–34.

CNN (1998) "Famine May Have Killed 2 Million in North Korea," CNN.com, August 19, 1998. www.cnn.com/WORLD/asiapcf/9808/19/nkorea.famine/. Accessed August 24, 2004.

CNN (2002) "Israeli report links Kenya terrorist to al Qaeda," CNN.com, November 29, 2002. www.cnn.com/2002/WORLD/africa/11/28/kenya.israel/. Accessed August 26, 2004.

CNN (2003) "Key Riyadh bombing suspect gives up," CNN.com, June 27, 2003. www.cnn.com/2003/WORLD/meast/06/26/riyadh.bombing.suspect/. Accessed August 24, 2004.

CNN (2004) "Bomb Kills Chechen President," CNN.com, May 9, 2004. www.cnn.com/2004/WORLD/europe/05/09/grozy.blast/index.html. Accessed July 2, 2004.

CNN (2005) "Army Reserve chief raises concerns about 'broken force'," CNN.com, June 1, 2005. www.cnn.usnews.printthis.clickability.com/pt/cpt?action+cpt&title. Accessed September 1, 2005.

Cohen, S. (1963) *Geography and Politics in a World Divided*, New York: Random House.

Cohen, S.B. (2002) "Some Afterthoughts: A Cohen Perspective from Mid-October 2001," *Political Geography* 21, pp. 569–72.

Collins, J.L. (1969) *War in Peacetime*, Boston, MA: Houghton Mifflin Company.

Cox, K.R. (2002) *Political Geography: Territory, State, and Society*, Oxford: Blackwell Publishers.

Crenshaw, M. (1981) "Thoughts on Relating Terrorism to Historical Contexts" in Crenshaw, M. (ed.) *Terrorism in Context*, University Park, PA: Pennsylvania State University Press, pp. 3–24.

Cresswell, T. (1996) *In Place/Out of Place*. Minneapolis, MN: University of Minnesota Press.

Crossette, B. (1998) "Violation; An Old Scourge of War Becomes Its Latest Crime," *The New York Times* July 14, 1998.

Cullison, Alan (2002) "Russia's Left-wing Politicians Retreat from Their Support of US-led War," *Wall Street Journal* February 5, 2002.

Cumings, B. (1997) *Korea's Place in the Sun*, New York: W.W. Norton & Company.

Dahlman, C. (2005) "Geographies of Genocide and Ethnic Cleansing: The Lessons of Bosnia-Herzegovina" in Flint, C. (ed.) *The Geography of War and Peace*, Oxford: Oxford University Press, pp. 174–97.

Dalby, S. (2003) "Geopolitics, the Bush Doctrine, and War on Iraq," *The Arab World Geographer* 6, pp. 7–18.

DARPA (2005) "'Falcon' Defense Advanced Research Projects Agency," January 7, 2005. www.darpa.mil/tto/programs/falcon.html. Accessed January 31, 2005.

Der Derian, J. (2001) *Virtuous War: Mapping the Military-Industrial-Media-Entertainment Network*, Boulder, CO: Westview Press.

Dewan, R. (2002) "What Does Azadi Mean To You?" in Butalia, U. (ed.) *Speaking Peace: Women's Voices from Kashmir*, London: Zed Books, pp. 149–61.

Dodds, K. (2003) "Licensed to Stereotype: Geopolitics, James Bond, and the Spectre of Balkanism," *Geopolitics* 8, pp. 125–56.

Donnan, H. and Wilson, T.M. (1999) *Borders: Frontiers of Identity, Nation and State*, Oxford: Berg.

Dowler, L. (2005) "Amazonian Landscapes: Gender, War, and Historical Repetition" in Flint, C. (ed.) *The Geography of War and Peace*, Oxford: Oxford University Press, pp. 133–48.

Drysdale, A. and Blake, G.H. (1985) *The Middle East and North Africa: A Political Geography*, New York: Oxford University Press.

Eckert, C.J., Lee, K.-B., Lew, Y., Robinson, M., and Wagner, E.W. (1990) *Korea Old and New*, Seoul: Ilchokak Publishers.

The Economist (2004) "Plots, Alarms, and Arrests," August 14, 2004, pp. 22–4.

Eksteins, M. (1989) *Rites of Spring: The Great War and the Birth of the Modern Age*, New York: Anchor Books.

Enloe, C. (1983) *Does Khaki Become You? The Militarisation of Women's Lives*, London: Pluto Press.

Enloe, C. (1990) *Bananas, Beaches, and Bases: Making Feminist Sense of International Politics*, Berkeley, CA: University of California Press.

Enloe, C. (2004) *The Curious Feminist: Searching for Women in a New Age of Empire*, Berkeley, CA: University of California Press.

Etzold, T.W. and Gladdis, J.L. (eds) (1972) *Containment: Documents on American Policy and Strategy 1943–1950*. New York: Columbia University Press.

Falah, G.-W. (2005) "Peace, Deception, and Justification for Territorial Claims: The Case of Israel" in Flint, C. (ed.) *The Geography of War and Peace*, Oxford: Oxford University Press, pp. 297–320.

Falah, G.-W. and Flint, C. (2004) "Geopolitical Spaces: The Dialectic of Public and Private Space in the Palestine-Israel Conflict," *Arab World Geographer* 7, pp. 117–34.

Flint, C. (2001) "The Geopolitics of Laughter and Forgetting: A World-systems Interpretation of the Post-modern Geopolitical Condition," *Geopolitics* 6, pp. 1–16.

Flint, C. (2003a) "Terrorism and Counterterrorism: Geographic Research Questions and Agendas," *The Professional Geographer* 55, pp. 161–9.

Flint, C. (2003b) "Geographies of Inclusion/Exclusion" in Cutter, S.L., Richardson, D.B., and Wilbanks, T.J. (eds) *The Geographical Dimensions of Terrorism*, New York: Routledge.

Flint, C. (2004) "The 'War on Terrorism' and the 'Hegemonic Dilemma': Extraterritoriality, Reterritorialization, and the Implications for Globalization" in O'Loughlin, J., Staeheli, L., and Greenberg, E. (eds) *Globalization and Its Outcomes*, New York: The Guilford Press, pp. 361–85.

Flint, C. (2005) "Dynamic Metageographies of Terrorism: The Spatial Challenges of Religious Terrorism and the 'War on Terrorism'" in Flint, C. (ed.) *The Geography of War and Peace*, Oxford: Oxford University Press, pp. 198–216.

Flint, C. and Falah, G.-W. (2004) "How the United States Justified its War on Terrorism: Prime Morality and the Construction of a 'Just War,'" *Third World Quarterly* 25, pp. 1379–99.

The Fog of War: An Errol Morris Film (2003) Sony Pictures Classic.

Friedman, T.L. (1995) *From Beirut to Jerusalem*, New York: Anchor Books.

Fromkin, D. (1989) *A Peace to End All Peace: The Fall of the Ottoman Empire and the Creation of the Modern Middle East*, New York: Owl Books.

Fukuyama, F. (1992) *The End of History and the Last Man*, New York: The Free Press.

Fussell, P. (1989) *Wartime: Understanding and Behavior in the Second World War*, Oxford: Oxford University Press.

Garamone, J. (2004) "Global Posture Part And Parcel of Transformation," Defenselink News, October 14, 2004. www.defenselink.mil/news/Oct2004/n10142004_2004101410.html. Accessed October 15, 2004.

Gilmartin, M. and Kofman, E. (2004) "Critically Feminist Geopolitics" in Staeheli, L.A., Kofman, E., and Peake, L.J. (eds) *Mapping Women, Making Politics*, New York and London: Routledge, pp. 113–25.

Glassner, M. and Fahrer, C. (2004) *Political Geography*, third edition, Hoboken: Wiley & Sons.

Gramsci, A. (1971) *Selections from Prison Notebooks*, London: Lawrence & Wishart.

Gregory, D. (2004) *The Colonial Present*, Oxford: Blackwell Publishing.

Griffin, M. (2003) *Reaping the Whirlwind: Afghanistan, al Qa'ida and the Holy War*, London: Pluto Press.

Gunaratna, R. (2002) *Inside al Qaeda: Global Network of Terror*, London: Hurst & Company.

Gurr, T.R. (2000) *Peoples Against States: Minorities at Risk in the New Century*, Washington, DC: United States Institute of Peace Press.

Gusev, D. (1996) "Vladimir Zhirinovsky," October 17, 1996. www.cs.indiana.edu/~dmiguse/Russian/vzbio.html. Accessed July 29, 2004.

Haggett, P. (1979) *Geography: A Modern Synthesis*, third edition, New York: Harper & Row.

Halliday, F. (1983) *The Making of the Second World War*, London: Verso.

Haraway, D. (1998) "Situated Knowledges: The Science Question in Feminism and the Privilege of Partial Perspective," *Feminist Studies* 14, pp. 575–99.

Hardt, M. and Negri, A. (2000) *Empire*, Cambridge, MA: Harvard University Press.

Hass, A. (2000) *Drinking the Sea at Gaza*, New York: Owl Books.

Haushofer, K., Obst, E., Lautensach, H., and Maull, O. (1928) *Bausteine zur Geopolitik*, Berlin: Kurt Vowinckel Verlag.

Hedges, C. (2003) *War is a Force That Gives Us Meaning*, New York: Anchor Books.

Henrikson, A.K. (2005) "The Geography of Diplomacy" in Flint, C. (ed.) *The Geography of War and Peace*, Oxford: Oxford University Press, pp. 369–94.

Heo, U. and Hyun, C.-M. (2003) "The 'Sunshine' Policy Revisited" in Heo, U. and Horowitz, S.A. (eds) *Conflict in Asia*, Westport, CT: Praeger.

Herb, G.H. and Kaplan, D.H. (1999) *Nested Identities*, Lanham, MD: Rowman & Littlefield.

Herbst, J.I. (2000) *States and Power in Africa: Comparative Lessons in Authority and Control*, Princeton, NJ: Princeton University Press.

Herod, A. and Wright, M.W. (2002) *Geographies of Power: Placing Scale*, Oxford: Blackwell Publishers.

Herzl, T. (2004) *Der Judenstad: Versuch einer modernen Lösung der Judenfräge: Texte und Materialen 1896 bis heute: herausgegeben und mit einem Nachwort von Ernst Piper*. Berlin: Philo.

Hoare, J. and Pares, S. (1999) *Conflict in Korea*, Denver, CO: ABC-CLIO.

Hoffman, B. (1998) *Inside Terrorism*, New York: Columbia University Press.

Hoffman, B. (2002) "Lessons of 9/11," Joint Inquiry Staff Request, October 8, 2002.

Huang, R. (2002) "Al Qaeda in Southeast Asia: Evidence and Response," Center for Defense Information, February 8, 2002, www.cdi.org/terrorism/sea.cfm. Accessed August 20, 2004.

Hubbard, P., Kitchin, R., Bartley, B., and Fuller, D. (2002) *Thinking Geographically: Space, Theory and Contemporary Human Geography*, London: Continuum.

Human Rights Watch (1999) "The Aftermath: The Ongoing Issues Facing Kosovar Albanian Women," www.hrw.org/reports/2000/fry/Kosov003.htm#P38_1195. Accessed July 22, 2004.

Human Rights Watch (2004) "On the Situation of Ethnic Chechens in Moscow," February 24, 2004. www.hrw.org/backgrounder/eca/russia032003.htm. Accessed July 7, 2004.

Huntingdon, S.P. (1993) "The Clash of Civilizations?" *Foreign Affairs* 72, pp. 22–49.

Islamic Republic of Pakistan (2004) "Frequently Asked Questions," May 25, 2004 www.infopak.gov.pk/public/kashmir/q&a.htm#didnot. Accessed June 2, 2004.

James, L. (1994) *The Rise and Fall of the British Empire*, New York: St Martin's Press.

Johnston, R.J. and Sidaway, J.D. (2004) *Geography and Geographers: Anglo-American Human Geography since 1945*, sixth edition, London: Arnold.

Kane, T. (2004) *Troop Deployment Dataset, 1950–2003*, The Heritage Foundation, Centre for Data Analysis. October, 2004. www.heritage.org/Research/NationalSecurity/troopsdb. cfm. Accessed November 17, 2004.

Kaplan, R. (1994) "The Coming Anarchy," *Atlantic Monthly* 273, pp. 44–76.

Kapuscinski, R. (1992) *The Soccer War*, New York: Vintage Books.

Kennedy, P. (1988) *The Rise and Fall of Great Powers*, New York: Random House.

Khashan, H. (2000) *Arabs at the Crossroads: Political Identity and Nationalism*, Gainesville, FL: University of Florida Press.

Klare, M. (1996) *Rogue States and Nuclear Outlaws: America's Search for a New Foreign Policy*, New York: Hill & Wang.

Klare, M. (2004) *Blood and Oil*, New York: Metropolitan Books.

Knox, P.L. and Marston, S. (1998) *Places and Regions in Global Context: Human Geography*, Upper Saddle River, NJ: Prentice-Hall.

Konrád, G. (1984) *Antipolitics*, San Diego, CA: Harcourt, Brace, and Jovanovich.

LaFraniere, S. (2003) "Rivalry Fragments Russia's Liberal," *The Washington Post* July 2, 2003. www.ncsj.org/AuxPages/020703WPost.shtml. Accessed July 27, 2004.

Laqueur, W. (1987) *The Age of Terrorism*, Boston, MA: Little, Brown, & Company.

Lenin, V.I. (1939) *Imperialism: The Highest Stage of Capitalism*, New York: International Publishers.

Lewis, B. (2002) *What Went Wrong? Western Impact and Middle Eastern Response*, Oxford: Oxford University Press.

Lewis, M.W. and Wigen, K.E. (1997) *The Myth of Continents: A Critique of Metageography*, Berkeley, CA: University of California Press.

Lineback, N.G. (2002) "India's Religious Quarrels," Appalachian State University. www.geo.appstate.edu/616_032202indiac.pdf. Accessed June 9, 2004.

Long, J.M. (2004) *Saddam's War of Words*, Austin, TX: University of Texas Press.

McAlister, M. (2001) *Epic Encounters: Culture, Media, and US Interests in the Middle East, 1945–2000*, Berkeley, CA: University of California Press.

Mace, J.E. (1984) "The Famine: Stalin Proposes a 'final solution'," *The Ukraine Weekly* June 24, 1984.

Malik, I. (2002) *Kashmir: Ethnic Conflict International Dispute*, Oxford: Oxford University Press.

Mansfield, P. (1992) *The Arabs*, third edition, London and New York: Penguin.

Martínez, O.J. (1994) *Border People: Life and Society in the US-Mexico Borderlands*, Tucson, AZ: University of Arizona Press.

Massey, D. (1994) *Space, Place, and Gender*, Minneapolis, MN: University of Minnesota Press.

Modelski, G. (1987) *Long Cycles in World Politics*, Seattle, WA: University of Washington Press.

Modelski, G. and Thompson, W. (1995) *Leading Sectors and World Powers*, Columbia, SC: University of South Carolina Press.

Mosse, G.L. (1975) *The Nationalization of the Masses*, New York: H. Fertig.

National Geospatial-Intelligence Agency (2004) *Geospatial Intelligence (GEOINT) Basic Doctrine*, Washington, DC: National Geospatial-Intelligence Agency.

NSS (National Security Strategy of the United States of America) (2002) Washington, DC: The White House.

Netanyahu, B. (2000) *A Durable Peace: Israel and its Place among the Nations*, New York: Warner Books.

Newman, D. (2005) "Conflict at the Interface: The Impact of Boundaries and Borders on Contemporary Ethnonational Conflict" in Flint, C. (ed.) *The Geography of War and Peace*, Oxford: Oxford University Press, pp. 321–44.

NSC-68: United States Objectives and Programs for National Security (1950) April 7, 1950. www.fas.org/irp/offdocs/nsc-hst/nsc-68.htm. Accessed July 22, 2004.

Oas, I. (2005) "Shifting the Iron Curtain of Kantian Peace: NATO Expansion and the Modern Magyars" in Flint, C. (ed.) *The Geography of War and Peace*, Oxford: Oxford University Press, pp. 395–414.

Oberdorfer, D. (2001) *The Two Koreas*, Indianapolis. IN: Basic Books.

O'Lear, S. (2004) "Resources and Conflict in the Caspian Sea," *Geopolitics* 9, pp. 161–86.

O'Loughlin, J. (ed.) (1994) *Dictionary of Geopolitics*, Westport, CT: Greenwood Press.

Ó Tuathail, G. (1996) *Critical Geopolitics*, Minneapolis, MN: University of Minnesota Press.

Painter, J. (1995) *Politics, Geography and "Political Geography,"* London: Arnold.

Parker, G. (1985) *Western Geopolitical Thought in the Twentieth Century*, London: Croom Helm.

Peet, R. (1998) *Modern Geographical Thought*, Oxford: Blackwell Publishers.

Peltz, E., Halliday, J.M., and Hartman, S.L. (2003) *Combat Service Support Transformation: Emerging Strategies for Making the Power Projection Army a Reality*, Santa Monica, CA: RAND.

Podur, J. (2002) "Kashmir Timeline," January 10, 2002. www.zmag.org/southasia/kashtime. htm. Accessed November 1, 2004.

Polgreen, L. (2005) "Darfur's Babies of Rape are on Trial from Birth," *The New York Times*, February 11, 2005, p. A1.

Prescott, J.R.V. (1987) *Political Frontiers and Boundaries*, London: Unwin Hyman.

Priest, D. (2003) *The Mission: Waging War and Keeping Peace with America's Military*, New York: W.W. Norton and Company.

Quadrennial Defense Review Report (2001) Washington, DC: Department of Defense.

Raina, S.B. (2002) "Leaving Home," in Butalia, U. (ed.) *Speaking Peace: Women's Voices from Kashmir*, London: Zed Books, pp. 178–84.

Ramet, S.P. (1999) *Gender Politics in the Western Balkans*, University Park, PA: Pennsylvania State University Press.

Ranstorp, M. (1998) "Interpreting the Broader Context and Meaning of Bin Ladens *Fatwa*," *Studies in Conflict and Terrorism* 21, pp. 321–30.

Rapoport, D.C. (2001) "The Fourth Wave: September 11 in the History of Terrorism," *Current History* 100, pp. 419–24.

Rose, G. (1997) "Situating Knowledges: Positionality, Reflexivities and Other Tactics," *Progress in Human Geography* 21, pp. 305–20.

Said, E. (1979) *Orientalism*, New York: Vintage Books.

Said, E. (2004) *From Oslo to Iraq and the Roadmap*, London: Bloomsbury.

Sankalp, S. (2003) "Spotlight," Indian National Congress, July 9, 2003. www.indiannational congress.org/shimla_sankalp.shtml. Accessed May 30, 2004.

Santhanam, K., Sreedhar, Saxena, S. and Manish (2003) *Jihadis in Jammu and Kashmir*, New Delhi: Sage Publications.

Schmitt, E. (2005) "Rumsfeld Warns of Concern About Expansion of China's Navy," *The New York Times*, February 18, 2005, p. A6.

Schofield, C., Newman, D., Drysdale, A., and Brown, J.A. (eds) (2002) *The Razor's Edge: International Boundaries and Political Geography*, London: Kluwer Law International.

Schofield, R. (2003) "Britain and Kuwait's Borders, 1902–23" in Slot, B.J. (ed.) *Kuwait: The Growth of a Historic Identity*, London: Arabian Publishing, pp. 58–94.

Semple, K. (2005) "Part-time Warriors Find Themselves Fully Committed," *The New York Times*, February 14, 2005, p. A2.

Sharp, J. (2000a) *Condensing the Cold War:* Reader's Digest *and American Identity*, Minneapolis, MN: University of Minnesota Press.

Sharp, J. (2000b) "Refiguring Geopolitics: The *Reader's Digest* and Popular Geographies of Danger at the End of the Cold War" in Dodds, K. and Atkinson, D. (eds) *Geopolitical Traditions: A Century of Geopolitical Thought*, London and New York: Routledge, pp. 332–52.

Shlaim, A. (2001) *The Iron Wall: Israel and the Arab World*, New York: W.W. Norton and Company.

Slot, B.J. (2003) "Kuwait: the Growth of a Historic Identity" in Slot, B.J. (ed.) *Kuwait: The Growth of a Historic Identity*, London: Arabian Publishing, pp. 5–29.

Smith, A.D. (1993) *National Identity*, Reno, NV: University of Nevada Press.

Smith, A.D. (1995) *Nations and Nationalism in a Global Era*, Cambridge: Polity Press.

Smith, N. (2003) *American Empire: Roosevelt's Geographer and the Prelude to Globalization*, Berkeley, CA: University of California Press.

Smith, W.D. (1986) *The Ideological Origins of Nazi Imperialism*, Oxford: Oxford University Press.

Sorokin, P.A. (1937) *Social and Cultural Dynamics, Volume 3*, New York: American Book Company.

Srivastava, M. (2001) *International Dimensions of Ethnic Conflict: A Case Study of Kashmir and Northern Ireland*, New Delhi: Bhavana Books and Print.

Staeheli, L.A. and Kofman, E. (2004) "Mapping Gender, Making Politics: Toward Feminist Political Geographies" in Staeheli, L.A., Kofman, E., and Peake, L.J. (eds) *Mapping Women, Making Politics*, New York and London: Routledge, pp. 1–13.

Staeheli, L.A., Kofman, E., and Peake, L.J. (eds) (2004) *Mapping Women, Making Politics*, New York and London: Routledge.

Stallworthy, J. (ed.) (1983) *Wilfred Owen: The Complete Poems and Fragments*. Oxford: Chatto and Windus, The Hogarth Press and Oxford University Press.

State of the Union Speech (2002) January 29, 2002. www.whitehouse.gov/news/releases/2002/01/print/20020129–11.html. Accessed July 26, 2004.

Stephanson, A. (1989) *Kennan and the Art of Foreign Policy*. Boston, MA: Harvard University Press.

Stump, R.W. (2005) "Religion and the Geographies of War" in Flint, C. (ed.) *The Geography of War and Peace*, Oxford: Oxford University Press, pp. 144–73.

Sumartojo, R. (2004) "Contesting Place: Anti-gay and Lesbian Hate Crime in Columbus, Ohio" in Flint, C. (ed.) *Spaces of Hate*, New York and London: Routledge, pp. 87–107.

Taylor, P.J. (1990) *Britain and the Cold War: 1945 as Geopolitical Transition*, London: Pinter Publishers.

Taylor, P.J. and Flint, C. (2000) *Political Geography: World-economy, Nation-state, and Locality*, fourth edition, Harlow: Prentice Hall.

Thompson, E.P. (1985) *The Heavy Dancers*, London: Merlin Press.

Tuchman, B. (1962) *The Guns of August*, New York: Dell Publishing.

Upadhyay, R. (2000) "The Muslim Agenda of the BJP—Will it work?" South Asia Analysis Group, September 28, 2000. www.saag.org/papers2/paper149.html. Accessed June 4, 2004.

US Census Bureau (2005) http://quickfacts.census.gov.qfd. Accessed January 31, 2006.

US CIA (2006) "The Caucasus and Central Asia," Perry Castaneda Library Map Collection, University of Texas. www.lib.utexas.edu/maps/commonwealth/caucasus_cntrl_asia_pol_00.jpg. Accessed 6 March, 2006.

Walsh, N.P. (2004) "'A Second Chechnya.' Ingushetia Dispatch," June 18, 2004. www.guardian.co.uk/elsewhere/journalist/story/0,7792,1242054,00.html. Accessed July 14, 2004.

Women's Initiative (2002) "Women's Testimonies fom Kashmir," in Butalia, U. (ed.) *Speaking Peace: Women's Voices from Kashmir*, London: Zed Books, pp. 82–95.

Women's Organization from Burma, Women's Affairs Department and National Coalition Government of the Union of Burma (2000) "Burma: The Current State of Women in Conflict Areas," January 2000. www.earthrights.org/women/ShadowWRP.doc. Accessed July 20, 2004.

Wong, E. (2005) "Iraqi Kurds Detail Their Strong Demands for Autonomy," *The New York Times*, February 18, 2005, p. A8.

World Bank (2005) http://web.worldbank.org/WBSITE/EXTERNAL/DATASTATISTICS/0,,contentMDK:20535285~menuPK:1192694~pagePK:64133150~piPK:64133175~theSitePK:239419,00.html. Accessed January 31, 2006.

World Oil (2003) "Greatest Oil Reserves by Country," *World Oil* 224.

INDEX

Page references in **bold** indicate figures; page references in *italic* indicate activities, boxes and tables.

Bush, President George W.: Blair's support for 55, 58; deployment of US troops in Central Asia 63; geopolitics portrayed in *Fahrenheit 9/11* 80; justification of invasion of Iraq 71, 79, 80–1, 81, 101, 182, 185; rhetoric of boundary defense 154; and similarities of Truman's foreign policy 67; State of the Union Speech (2002) 74, 121, 153

Call, Chuck *167*
Camp David peace agreement 141–2
"capability threat" 47
capitalism: landscape of Moscow 9, **9**; view of Iraq war in *Fahrenheit 9/11* 80; Wallerstein's world-system theory *35*
Caribbean 59, 67
Carter, President Jimmy *76*
Caspian Sea basin: oil and gas reserves 61, **62**, 63
Caucacus region 61, 119, **134**
CDI Russia Weekly 63, 64
Central America *167*
Central Asia: decoding geopolitics of 61–4; deportations of Chechens and Ingushes to 113; recent US deployment of troops to 63, *210*; Russia's power in 119; US military aid to republics 59; and "War on Terrorism" 13, 63
Central Europe: and European Union 41; police states during Cold War 166; renunciation of Communist system 40; US bases in 47
Chechens: deportations by Stalin 113; perspective of conflict with Russia 119; plight of refugees 116–17, 118
Chechnya **117**; attempts to justify invasion of 117–18, 185; conflict with Russia 112–13, 113–19
Chhachhi, S. 205, 207, 208
Chile 4, *167*
China: contemporary geopolitical code *76*; economic growth and energy demands 61, 208–9, 211; expanding influence in Southeast Asia 33, *34*, 58, 66; expansion of navy 209; geopolitical code regarding Central Asia 61, 63–4; increasing political and economic power 33–4; and Korean War 151; as possible threat to US 50; territorial claims in East and Southeast Asia 126; territorial dispute with India 200, 201
Chinese community: Vancouver 8

Chinese Empire 126, 149
Christianity: and creation of American frontier 132; and "duty" of war 66
churches: British memorialization of war losses 66
Churchill, Prime Minister Winston: "Iron Curtain" speech *14*
CIA (Central Intelligence Agency) *60*
civilization: and comparison of US to classical societies 88; equated with US global mission 100; Huntington's ideas 92; mission of British Empire 66
classic geopolitics (end of nineteenth century) 17, *17*, 18, 158; and foreign affairs 28, 29; geodeterminism *22*; Germany 20–1; Modelski 36, 50–1; and situated knowledge 24, 33
Clinton, President Bill 153
CNN (Cable News Network) 93, 116, 179
Cohen, Saul 23
Cohen, S.B. 157
Cold War: and division of Korean peninsula 153; geopolitics 13, 86, 166; and global influence of Soviet Union 59; influence of classic geopoliticians on strategists 18, 20; interpreted within Modelski's model 40–1; "iron curtain" *14*; portrayal by *Reader's Digest* 89; Soviet control of Caspian Sea basin and Central Asia 61; strategic interest in Middle East 141; US "victory" 43–4
collective identity: and Darfur conflict 194; positive interpretations of nationalism 107, 128
Collective Security Treaty Organization (CSTO) 64
Colombia *167*
colonialism: Arab nationalist resistance to 95; and creation of boundaries in Africa 132; in Germany's geopolitical code 66; Saddam's rhetoric against 98; "scramble for Africa" 13; terrorism against 168
Columbus, Ohio: anti-gay hate crimes 7
combat 123, 124
Committee of Soldiers' Mothers (Russia) 118
"common sense" 85, 212
Communism: portrayal by *Reader's Digest* 88–9; Soviet Union 9; as "threat" during Cold War 13, **14**, 40, 87; US mission against 22, 67
computer games 93
conflicts: border disputes 132; contested places 10; structure and agents 30